GLENCOE
MATHEMATICS

MathScape

SEEING AND THINKING MATHEMATICALLY

Course 2

 **Glencoe**

New York, New York   Columbus, Ohio   Chicago, Illinois   Peoria, Illinois   Woodland Hills, California

 Glencoe

The *McGraw-Hill* Companies

Send all inquiries to:
Glencoe/McGraw-Hill
8787 Orion Place
Columbus, OH 43240

ISBN: 0-07-860467-2

5 6 111/058 08 07

TABLE OF CONTENTS

Buyer Beware 2
Rates, Ratios, Percents, and Proportions

Chance Encounters 46
Probability in Games and Simulations

Making Mathematical Arguments ...92
Generalizing About Numbers

From the Ground Up 136
Modeling, Measuring, and Constructing Homes

The Language of Algebra 180
Equations, Tables, and Graphs

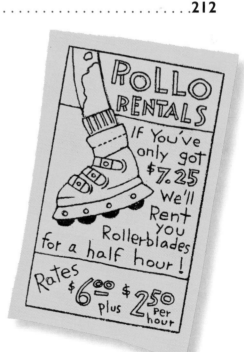

Getting Down to Business 224
Functions and Spreadsheets

Getting in Shape 270
Exploring Two-Dimensional Figures

Buyer Beware

For best buys...

How to:

- **Compare Cost**
- **Compare Quantities**
- **Save with Coupons**

Also...

Budget for a Banquet

Estimate Expenses

To: Staff Reporters
From: Buyer Beware Magazine, Inc.

Welcome to Buyer Beware. As a staff reporter, your job is to help your readers become more educated consumers. Mathematics is an important part of consumer awareness. You will use unit pricing to find the better buy, write ratios to compare brands, and use proportions to increase recipes and determine cost. You will use percents to estimate discounts and interpret data in circle graphs.

What math is involved in being an educated consumer?

BUYER
BEWARE

PHASE**ONE**
Rates

In this phase, you will compare the prices and sizes of products to determine which is the better buy. Next you will use a price graph to find and compare unit prices. Then you will decide when it is cheaper to buy by the pound. Finally, you will test claims made by a sandwich bar to see whether it is less expensive to buy a sandwich there or to make one at home.

PHASE**TWO**
Ratio and Proportion

You will begin this phase by writing ratios to compare quantities. Then you will use a ratio table to find equal ratios. Next, you will use equal ratios to compare brands. Then you will use proportions to solve problems in a variety of situations and determine what makes a situation proportional. You will end the phase by using what you know about ratio and proportion to create a mosaic design.

PHASE**THREE**
Percents

You will begin this final phase by using familiar benchmarks, then counting and rounding to estimate expenses in a budget. Next, you will interpret and create circle graphs representing budgets for a drama club. Then you will use percents and discount coupons to find your savings in a percent-off sale. You will end the unit through a final activity: planning an athletic banquet on a budget.

PHASE ONE

To: Staff Reporters
From: The Editors

Your first assignment is to test claims made about unit price. Super Sandwich Bar claims that it is less expensive to make a sandwich at their sandwich bar than it is to make one at home. When you complete the assignment we would like you to write an article for *Buyer Beware* stating your findings.

To prepare for this assignment, you will need some experience working with rates and unit prices.

Smart consumers want the best buy they can get for their money. To get the best buy, however, a consumer needs to look beyond advertised claims.

In order to do accurate comparison shopping, you need to know how to calculate unit price. By finding the unit price—the price per ounce, pound, or item—you will learn to make informed decisions and find the products that are better buys.

Rates

WHAT'S THE MATH?

Investigations in this section focus on:

DATA and STATISTICS

- Finding unit prices using a price graph
- Constructing a price graph to compare unit prices

NUMBER

- Comparing unit prices of different-size packages
- Comparing unit prices of different brands
- Finding the price per pound to decide the better buy
- Comparing prices per pound
- Calculating long-term savings
- Calculating unit prices and total prices

MathScape Online
mathscape2.com/self_check_quiz

What's the Best Buy?

Shoppers need to be able to calculate unit prices to find the best buy. In this lesson, you will compare various-size packages of cookies made by the same company to decide which size is the best buy. Then, you will compare two different brands of chocolate chip cookies to decide which one gives you more cookie for your money.

Compare the Unit Prices of Different-Size Packages

How can you compare cookie packages to find the best buy?

The Buyer Beware Consumer Research Group has collected data on chocolate chip cookies. They want you to find out which package of Choco Chippies is the best buy.

To find the best buy, you need to find the unit price, or the price per cookie, for each size package. You can find the price of one cookie in a package if you know the total amount of cookies in the package and the price of the package.

1 Use a calculator to figure out the price per cookie for each package. Round your answers to the nearest cent.

2 Decide which package size is the best buy. Explain how you figured it out.

3 List the different Choco Chippies package sizes in order from best buy to worst buy.

Choco Chippies Prices		
Package Size	**Number of Cookies**	**Package Price**
Snack	4	$0.50
Regular	17	$1.39
Family	46	$3.99
Giant	72	$5.29

SNACK SIZE   REGULAR SIZE   FAMILY SIZE   GIANT SIZE

Compare the Unit Prices of Two Different Brands

At *Buyer Beware* magazine we frequently get letters from our readers asking questions about best buys. Here is one of the letters we received:

> Dear Buyer Beware,
>
> Help! My friend and I don't agree on which brand of cookies is the best buy. She's convinced that it's Mini Chips, but I'm sure it's Duffy's Delights. Which is really the better buy?
>
> The Cookie Muncher

How can you use unit price to determine which of two brands is the better buy?

The research group at *Buyer Beware* has put together Cookie Prices data for you to use.

1. Decide which unit you would use to compare the two different brands of cookie. Explain why you chose that unit.

2. Find the price per unit of each brand of cookies.

3. Compare the unit price of the two brands of cookies. Is one brand the better buy? If so, explain why. If not, explain why the brands are equally good buys.

Cookie Prices

Brand	Package Price	Number of Cookies	Package Weight
Mini Chips	$1.39	17	6 oz
Duffy's Delights	$2.29	10	11 oz

Determine the Better Buy

Write a response letter to The Cookie Muncher.

- Describe what you did to figure out the better buy.
- Give evidence to support your conclusions.
- Give some general tips for finding the best buy.

MINI CHIPS

DUFFY'S DELIGHT

hot **words** | unit price
rate

Homework

page 34

2 The Best Snack Bar Bargain

You can use a price graph to compare unit prices for different products. In this lesson you will use a price graph to determine the price at different quantities of a snack bar if you were paying by the ounce. Then you will construct a price graph to compare the prices of five different products.

Use a Price Graph to Find Unit Price

How can you use a price graph to estimate snack bar prices?

The graph below shows the prices for three different snack bars. The price of Mercury Bars is $1.00 for 2 oz. Jupiter Bars are $2.98 for 3.5 oz and Saturn Bars are $3.50 for 4.5 oz.

Each of the three dots on the graph shows the price and the number of ounces for one of the snack bars. Each line shows the price of different quantities of the snack bar at the same price per ounce.

1. Use the price graph to find the price of a 3-oz Mercury Bar.

2. Use the price graph to find the price of a 0.5-oz Saturn Bar.

3. Use the price graph to find which snack bar has the lowest unit price and which has the highest unit price.

4. Use your calculator to find the unit price of 1 oz of each of the snack bars. Use the price graph to check your calculations.

Snack Bar Line Graph

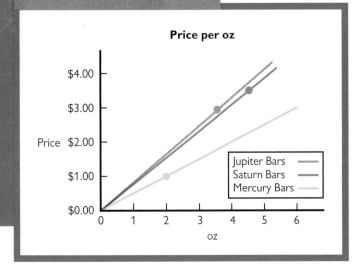

Construct a Price Graph to Compare Unit Prices

The Buyer Beware research group has collected data on five different products.

Product	Number of Units	Price
Oatmeal	14 oz	$2.40
Tuna	7.5 oz	$2.00
Penne pasta	8 oz	$1.00
Sourdough pretzels	12 oz	$2.60
Whole wheat rolls	9 oz	$1.60

1 Use the above data to construct a price graph.

2 Use the price graph to complete a Price Comparison. For each product, record the price for 1 oz, 3 oz, 4.5 oz, and 6 oz.

Price Comparison Table

Product	1 oz	3 oz	4.5 oz	6 oz
Oatmeal				

3 Which product is the most expensive per ounce? the least expensive per ounce?

Write About Your Price Graph

Think about what you have learned about interpreting and making price graphs.

- Write a description of your price graph and what it shows. Explain how to use your price graph to find prices for packages that are larger and smaller than your original package.

- Describe how two products will look on a price graph if their prices are almost the same per ounce.

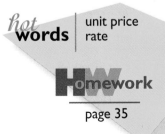

hot **words** | unit price
rate

Homework
page 35

3 Cheaper by the Pound

USING UNIT PRICES
AND WEIGHT UNITS
TO COMPARE PRICES

Sometimes buying in "bulk" or larger quantities will save you money. In this lesson you will find the price per pound of different-size packages of rice to decide which one is the best buy. Then you will evaluate the price per pound of items at a silly sale.

Find the Price per Pound to Decide the Best Buy

How can you find the package that is the least expensive per pound?

1 What is the price per pound of each package of rice? Round your answer to the nearest cent.

2 Find out which package of rice is the best buy in terms of price per pound.

3 Explain which package of rice makes the most sense to buy if only one person in a family eats rice.

4 Decide which package of rice would be the best buy for your family.

Fluffy Rice	
2 lbs	$1.09
5 lbs	$2.69
10 lbs	$4.99
20 lbs	$7.99

2 POUNDS
$1.09

5 POUNDS
$2.69

10 POUNDS
$4.99

20 POUNDS
$7.99

Compare Prices per Pound

Suppose you go to a silly sale where everything is rated by price per pound. Which is cheapest per pound: a bicycle, a pair of sneakers, a video camera, or a refrigerator?

How can you determine the prices per pound of different items?

43 POUNDS
$139.99

2 POUNDS
$84.99

11 POUNDS
$799.00

230 POUNDS
$679.99

1 Use your calculator to find the price per pound of each item.

2 Rank the items from the least to most expensive per pound.

3 What items have high prices per pound?

4 What items have low prices per pound?

Write About Buying in Bulk

Write what you know about buying in bulk. Be sure to answer these questions in your writing:

- How can you figure out if the largest size is the least expensive per pound?

- Is the largest size always the least expensive per pound?

- When is it a good idea to buy products in bulk? When is it not a good idea?

- Could the same purchase be a good choice for one consumer but not for another consumer? Explain.

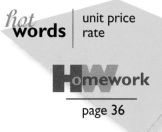

hot **words** | unit price
rate

Homework

page 36

4 It Really Adds Up

In this lesson you will use what you have learned about unit prices to solve real-life problems. First, you will find and compare the prices of two different brands of pretzels. Then you will take on the role of an investigative reporter to test the claim made by a sandwich bar.

Calculate Long-Term Savings

How much can you save over time by buying a less expensive brand?

People are often surprised at how buying a little snack every day can really add up over time. The Buyer Beware research team wants you to figure out the price of buying two brands of pretzels for different time periods.

1 Find the price of buying one bag of pretzels every day for a week, a month, and a year. Make a table like the one below to organize your answers.

2 How much would you save if you bought No-Ad Pretzels instead of Crunchy Pretzels for the different time periods?

3 If you bought a bag of No-Ad Pretzels every day instead of Crunchy Pretzels, how many days would it take to save $30? $75? Explain how you figured it out.

Pretzel Costs

Type of Pretzels	Price for I Bag a Day for I Day	Price for I Bag a Day for I Week	Price for I Bag a Day for I Month (4.3 weeks)	Price for I Bag a Day for I Year
Crunchy Pretzels	$0.65			
No-Ad Pretzels	$0.50			
Savings for buying No-Ad Pretzels				

Calculate Unit Prices and Total Prices

At Super Sandwich Bar, customers can make their own Terrific Ten Sandwich—two slices of cheese and eight slices of meat—for only $3.59. The restaurant claims that this is cheaper than making the sandwich at home. The Buyer Beware research group wants you to test this claim. They have collected information on the prices of different ingredients for you to use.

How can you test a claim by figuring out unit prices?

Sandwich Tips

STEPS TO DESIGNING A SUPER SANDWICH

1. Choose *at least* three ingredients from the handout In Search of the Terrific Ten Sandwich.

2. Make a table. List the ingredients you chose, the amount of each, and the price of each.

3. What is the total price of your giant sandwich if you paid for it by the slice?

4. Do you think Super Sandwich Bar's price of $3.59 is a good deal? Explain your reasoning.

5. Name your sandwich and draw a picture of it for a magazine article.

Write About Super Sandwich Bar's Claims

- Describe the strategies and solutions you used to figure out the total price of your sandwich.

- Write a magazine article discussing your findings about the claims made by Super Sandwich Bar. In your article answer the question: Is it cheaper to make a sandwich at the bar or buy all the ingredients and make it yourself?

hot **words** | unit price

HW**omework**

page 37

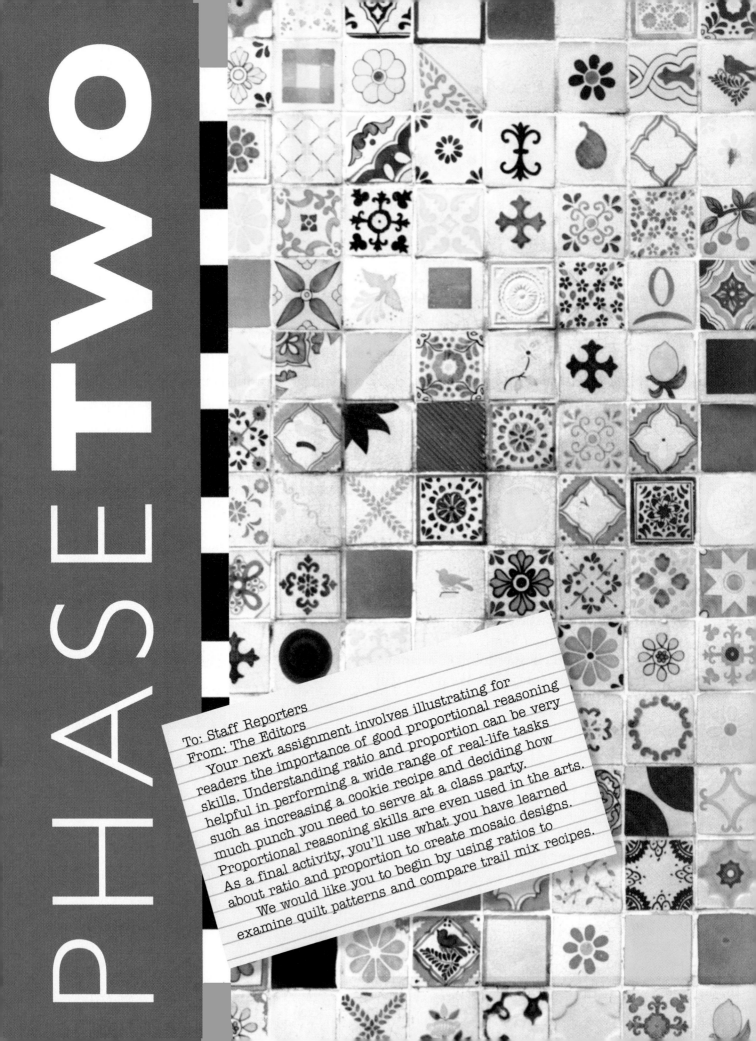

PHASE TWO

To: Staff Reporters
From: The Editors

Your next assignment involves illustrating for readers the importance of good proportional reasoning skills. Understanding ratio and proportion can be very helpful in performing a wide range of real-life tasks such as increasing a cookie recipe and deciding how much punch you need to serve at a class party. Proportional reasoning skills are even used in the arts.

As a final activity, you'll use what you have learned about ratio and proportion to create mosaic designs.

We would like you to begin by using ratios to examine quilt patterns and compare trail mix recipes.

Can ratio and proportion be useful in planning school activities? A ratio is a relationship between two quantities of the same measure. Ratios are useful when you want to compare data, decide quantities of servings, or compare different products.

A proportion states that two ratios are equal. Setting up and solving a proportion can help you increase recipes or create a design to certain specifications.

Ratio and Proportion

WHAT'S THE MATH?

Investigations in this section focus on:

NUMBER

- Using ratios to compare data
- Using equivalent fractions to compare ratios

SCALE and PROPORTION

- Finding equal ratios with a ratio table
- Using cross products to compare ratios
- Using equal ratios and cross products to solve proportions
- Writing and using proportions to solve problems
- Determining when to use proportions to solve a problem

MathScape Online

mathscape2.com/self_check_quiz

5 Quilting Ratios

WRITING AND COMPARING RATIOS

A ratio is a comparison of one number to another by division. In this lesson, you will use ratios to describe a design. Then you will use ratios to compare trail mix recipes.

Practice Writing Ratios

How can you use part-to-part and part-to-whole ratios to describe a design?

Leticia is designing a quilt pattern that is made up of squares. One row of the quilt is shown below.

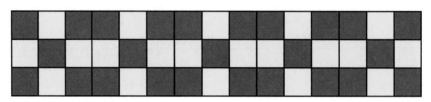

1. What is the ratio of blue squares to yellow squares? Is this a part-to-part or part-to-whole ratio?

2. What is the ratio of blue squares to all the squares in the row? Is this a part-to-part or part-to-whole ratio?

3. Gino is designing the quilted place mat at the right. Find two different ratios within the design. Describe them using words and numbers. Tell whether they are part-to-part or part-to-whole ratios.

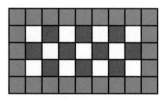

Writing Ratios

You can write ratios in several ways. For example, in one classroom, there are 2 boys for every 3 girls. This ratio can be written as follows.

$$2 \text{ to } 3 \quad 2:3 \quad \frac{2}{3}$$

These are part-to-part ratios. *There are 2 boys for every 3 girls in class.*

This class could also be described using a part-to-whole ratio. *There are 2 boys for every 5 students in class.*

Compare Ratios in Trail Mix

Dwight Middle School is planning an Autumn Festival. The seventh grade is planning to buy trail mix to sell at the festival. They want to decide if one of two brands, Mountain Trail Mix or Hiker's Trail Mix, contains more chocolate by weight or if both contain the same amount. The *Buyer Beware* research group has collected the information shown on the two brands. Note: The mixes contain only dried fruit and chocolate chips.

1 What is the ratio of chocolate chips to dried fruit by weight in Mountain Trail Mix?

2 What is the ratio of chocolate chips to dried fruit by weight in Hiker's Trail Mix?

3 Are the ratios of chocolate chips to dried fruit for each type of mix the same or different?

4 Does one trail mix brand contain more chocolate per ounce than the other? If so, which one? How do you know?

Mountain Trail Mix
6 oz chocolate chips
10 oz dried fruit

Hiker's Trail Mix
4 oz chocolate chips
6 oz dried fruit

How can comparing fractions help you decide which trail mix to buy?

Write Ratio Statements

Write five different ratio statements about Mountain Trail Mix. Here are two examples to help you get started.

> The ratio of chocolate chips to dried fruit is 6:10.
>
> There are 3 ounces of chocolate chips for every 5 ounces of dried fruit.

- For each ratio statement, label it either part-to-part or part-to-whole.

- Add one additional ingredient to the Mountain Trail Mix so that you can create either a part-to-part or a part-to-whole ratio of 2 to 3. How many ounces of the ingredient will you add? Possible ingredients include peanuts, coconut, pretzels, or cereal.

hot **words** | ratio
equivalent fractions

H W omework

page 38

6 In the Mix

USING RATIO TABLES

Ratio tables can help you organize information, identify a pattern and extend it. In this lesson, you will use ratio tables and other methods to create batches of fruit punch that maintain a consistent fruitiness.

Explore Methods for Increasing Quantities

How can you use ratios to increase recipes?

There are many kinds of drinks you can buy at the store in the form of a liquid concentrate to which you add water. For example, one kind of fruit punch uses 3 cans of water for each can of fruit punch concentrate.

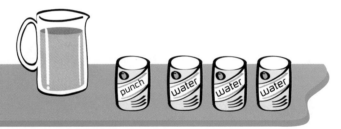

1 How many total cans of liquid would you use to make 2 batches of fruit punch? Each batch uses 1 can of fruit punch concentrate.

2 How would you create a bigger batch of punch that uses a total of 20 cans of liquid (both concentrate and water)? The punch should have the same amount of fruity taste as a single batch does. How many cans of fruit punch concentrate would you use? How many cans of water would you use?

Students are making punch for a class party and need to serve everyone who will be coming. From last year's party, the students estimate they will need to make 3 batches of punch.

3 How many cans of fruit punch concentrate will they need?

4 How many cans of water will they need?

Be ready to describe your thinking.

Compare Solutions

Three students explained how they created three batches of punch for the class party. Do you agree or disagree with each student's reasoning? If you disagree, explain why.

How can ratios help you solve a problem?

Kira

In one batch of the punch, 3 of the 4 cans are water, so $\frac{3}{4}$ of the punch is water. In my bigger batch, I'll need $\frac{3}{4}$ of the total to be water. If I triple the recipe, there will be a total of 3×4 or 12 cans of liquid in all. Since $\frac{3}{4}$ of 12 is 9, I'll use 9 cans of water and 3 cans of concentrate.

Here's Kira's picture.

Here's the bigger batch.

Tobi

The directions call for 1 can of concentrate and 3 cans of water. Since $3 - 1 = 2$, there are 2 more cans of water than cans of concentrate. In my bigger batch of punch, I'll need twelve cans in all. So I'll use 5 cans of concentrate and 7 cans of water.

Here's Tobi's picture.

Mitchell

I'll use a pattern to figure this out. I know I need 3 cans of water for each 1 can of concentrate. I'll just keep adding 1 can of concentrate and 3 cans of water until we have three batches.

hot words | ratio

Homework
page 39

Here's Mitchell's picture.

7 Halftime Refreshments

UNDERSTANDING PROPORTIONS

A proportion shows that two ratios are equal. In this lesson, you will use proportions to decide how many spoonfuls of hot cocoa mix are needed to make mugs of cocoa. Then you will use proportions to solve problems in other contexts.

Use Different Methods to Solve Proportions

How are proportions related to ratios?

Proportion

A proportion is a comparison between two equal ratios. Often, a proportion is written as two equivalent fractions.

For example, $\frac{12 \text{ inches}}{1 \text{ foot}} = \frac{36 \text{ inches}}{3 \text{ feet}}$.

What other proportions can you think of?

Set up a proportion and solve each problem below.

1 The seventh grade is planning to sell mugs full of hot cocoa at the football game. If 6 spoonfuls of cocoa mix make 3 mugs of hot cocoa, how many spoonfuls are needed to make 9 mugs?

2 How many spoonfuls would be needed for:

    a. 21 mugs?   **b.** 36 mugs?   **c.** 96 mugs?

3 How many mugs would you get from:

    a. 14 spoonfuls?   **b.** 30 spoonfuls?   **c.** 64 spoonfuls?

4 The seventh grade is going on a field trip to the local science museum. The school policy on field trips states that there must be one adult chaperone for every 8 students. How many chaperones are needed for:

    a. 40 students?   **b.** 96 students?   **c.** 120 students?

More Proportion Problems

Solve each problem.

How can you use what you know to solve problems using a proportion?

1 Owen is converting a cookie recipe to feed a large group of people. One of the ingredients his recipe calls for is 2 cups of flour. His recipe makes 3 dozen cookies. If he has 12 cups of flour, how many dozen cookies can he make?

2 Alecia is traveling in Canada with her family and notices that the road signs have distances in both miles and kilometers. At one point, the sign says the distance to Quebec is 65 miles or 105 kilometers. If their trip covers 250 kilometers in all, what distance will they have traveled, rounded to the nearest whole, in miles?

3 Jeanine is ordering pizzas to feed students who participate in the school car wash fund-raiser. At another school fund-raiser last month, 10 students ate 4 pizzas. How many pizzas should she plan to order if there will be 75 students participating?

4 Nurses sometimes use proportions when they take your pulse. A healthy heart rate is about 72 beats per minute. Some nurses take your pulse for 15 seconds, then estimate your heart rate. How many beats would they expect to count in 15 seconds if you have the average healthy heart rate?

Write About Proportions

Write your own problem that can be solved using proportions. Use the problems above as a model. Write the solution to the problem on a separate sheet of paper.

hot **words** | proportion
cross product

HW **omework**

page 40

 **Can I Use a Proportion?**

USING PROPORTIONS
TO SOLVE
PROBLEMS

In the lesson, you will examine real-life situations to determine whether they are proportional. Then, you will design a mosaic to show what you have learned about ratio and proportion.

Determine if a Situation is Proportional

How do you know when you can use a proportion to solve a problem?

Nellie's favorite trail mix has 4 ounces of peanuts and 6 ounces of chocolate. She wants to make a big batch with the same ratio of peanuts to chocolate to bring on a hike with her friends this weekend.

1 Complete the table to help Nellie make different sized batches.

Peanuts (ounces)	4	6	8	10	
Chocolate (ounces)	6				

2 On graph paper, plot each ordered pair from the table and draw a line through the points.

3 Is the ratio of peanuts to chocolate the same regardless of the size of the batch? Is this situation proportional? Explain.

Susan is going to a school fair. She has to pay $2 to enter the fair and $1 per ticket for games and refreshments.

4 Complete the table to help Susan determine her total expenses based on how many tickets she purchases.

Number of Tickets	4	6	8	10	
Cost (dollars)	6				

5 On graph paper, plot each ordered pair from the table and draw a line through the points.

6 Is the ratio of tickets purchased to total cost of the fair the same regardless of the number of tickets purchased? Is this situation proportional? Explain.

Write About Solving Proportions

A student designed the winning mosaic for the school's new entryway. Now the building committee needs to calculate the cost of her design given the following prices for color tiles.

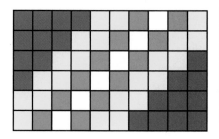

Blue tiles	5 for $4.00
Red tiles	5 for $6.00
Yellow tiles	5 for $3.00
White tiles	5 for $2.50

Refer to the handout Different Ways to Solve Proportions to help you answer each question.

1 Calculate the cost of blue, red, and yellow tiles in the mosaic design using a different method for each color. Write the name of the method you used and an explanation of how you calculated each cost using this method.

2 Calculate the total cost of the white tiles using any method you wish. Then write an explanation of how you calculated the cost using this method.

Design a Mosaic

In this activity, you will design a mosaic to show what you have learned about ratios and proportions.

Use the following guidelines to plan and create your design:

- Your design should have 3–6 different colors in it.

- Your design should have between 30–60 total squares of the same size in it.

- Two of your colors should be in a part-to-part ratio of 2:3.

- Some of your colors should be in a part-to-whole ratio of 1:3.

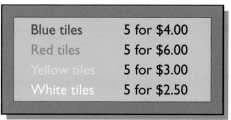

How can you use what you have learned to solve problems?

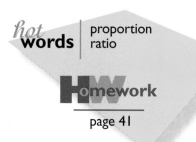

hot **words** | proportion ratio

Homework

page 41

PHASE THREE

To: Staff Reporters
From: The Editors

Your final assignment is to plan an athletic banquet on a budget. A budget is the amount of money that is available for a particular use. When you complete the assignment you will write a magazine article for *Buyer Beware* on "Planning a Banquet on a Budget."

To prepare for this assignment, you will need to get some experience working with percents.

In this final phase you will use percents to help you become a smart shopper. Knowing about percents will help you to interpret discounts and resulting sale prices. If you know how much the discount actually reduces the price of an item, you can make informed decisions about what to buy and what not to buy.

Percents

WHAT'S THE MATH?

Investigations in this section focus on:

NUMBER

- Estimating expenses
- Exploring counting and rounding
- Calculating discounts
- Following a budget
- Determining savings
- Estimating and calculating percents

DATA and STATISTICS

- Interpreting a circle graph

GEOMETRY and MEASUREMENT

- Constructing a circle graph

MathScape Online
mathscape2.com/self_check_quiz

Team Spirit

ESTIMATING PERCENTS

In this lesson you will use benchmarks and mental math to estimate the expenses of a football team. Then you will count and round to estimate the amount of money a field hockey team must raise to travel to a tournament.

How can you use benchmarks to help you estimate expenses?

Estimate Expenses

Estimating a percent of something is much the same as estimating a fractional part of something.

Last year the Cougars' football team's total expenses were $20,000. The table below shows what percent of the budget was spent on each expense. About how much money was spent on each item?

1 Make a table like the one shown below and complete the last column. Estimate the cost of each expense. Use benchmarks and mental math to make your estimates. Do not use paper and pencil.

2 The team spent exactly $800 on one expense. Which was it? Use estimation to figure it out.

3 The Tigers football team, cross-town rivals, spent 26% of its $16,000 budget on uniforms. Which team spent more on uniforms? Use estimation to figure it out.

Football Team Expenses

Items	Percent of Budget	Estimated Cost
Uniforms	23%	
Transportation	6%	
Coach's salary	48%	
Equipment	11%	
Officials' fees	4%	
Trainer's salary	8%	
Total Expenses:		**$20,000**

Explore Counting and Rounding

A girls' field hockey team wants to play in a tournament. The school district will pay only a certain percentage of each expense. The team must raise the amount not paid by the district.

If you can find 50%, 10%, and 1% of a number mentally, you can also estimate some other percents mentally. For example, think of 25% as half of 50% and 5% as half of 10%.

How can counting and rounding help you estimate expenses?

Field Hockey Team Trip Expenses

Expense	Estimated Cost	Percent District Will Pay
Transportation	$700	3%
Meals	$600	12%
Hotel	$925	4%
Tournament fees	$346	22%
Tournament uniforms	$473	2.5%

1 Use counting and rounding to estimate the dollar amount the school district will pay for each expense.

2 Estimate the total amount the district will pay for all the expenses.

3 Estimate the total amount of money the team will need to raise.

4 Use your calculator to figure out the exact amount the team will need to raise for each expense.

Write About Your Estimates

Write about the estimates you made for the field hockey team expenses.

- What strategies did you use to estimate the expenses and total cost?

- What are some ways to determine whether your estimate is reasonable?

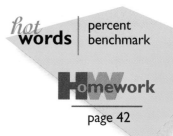

hot **words** | percent benchmark

HW**omework**

page 42

10 Playing Around

Circle graphs are useful tools for comparing percents.
They show how different parts are related to a whole. In this lesson you will use a calculator to analyze data in a circle graph. Then you will use survey information about fund-raisers to make your own circle graph.

Use a Calculator to Find Percents in a Circle Graph

How can you analyze data in a circle graph?

The drama club plans to present a production of Moss Hart's *You Can't Take It With You.* The cost of the production is displayed in the following circle graph.

1 What is the total cost of the production? Use the circle graph to estimate what percent of the total cost was spent on each expense. Tip: Compare the size of each section to the whole circle.

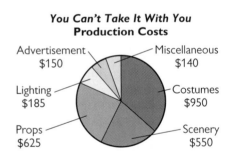

You Can't Take It With You Production Costs

Advertisement $150
Miscellaneous $140
Lighting $185
Costumes $950
Props $625
Scenery $550

2 Use your calculator to help you find the percent of each expense in the circle graph. First, find the percent that will be spent on costumes by expressing the amount as a fraction; for example:

$$\frac{\text{cost of costumes}}{\text{total cost of production}}$$

Then, change the fraction into a decimal. Use your calculator to help you with this. Divide the cost of the costumes by the total cost of the production. To express the decimal as a percent, multiply the decimal by 100, round to the nearest whole number, and add a percent (%) sign.

3 Compare your estimates with the exact percent you got using the calculator. How close were your estimates?

Construct a Circle Graph

The drama club decided it would like to attend a performance of Andrew Lloyd Webber's *Cats*. In order to see the musical production, they needed to raise the money to purchase the tickets. They conducted a survey in the middle school to find out which type of fund-raiser most students would be likely to attend. The survey gave the following results.

How can you construct a circle graph to show survey results?

Fund-Raiser Choices by Number of Students in Each Grade

Choice	6th Grade Students	7th Grade Students	8th Grade Students	Total Students	Percent of Students
Carnival	88	75	69		
Raffle	45	49	41		
Car wash	66	54	52		
Bake sale	34	32	33		
Candy sale	12	11	15		
Don't know	3	4	1		

1 Figure out the total number of students who were surveyed.

2 Use your calculator to find the percent of students in the middle school that chose each activity. Round your answers to the nearest whole number.

3 Follow the guidelines on the handout Circle Graph to help you construct a circle graph to display your data.

Conduct Your Own Survey

Conduct a survey in your math class to find out which fund-raiser your classmates would choose to attend.

- Figure out what percent of the class chose each type of fund-raiser.

- Make a circle graph to represent your class's survey results.

- Compare your class's results to the data in the table.

hot **words** | circle graph
angle

Homework

page 43

11 Sale Daze

USING PERCENTS
TO CALCULATE
SALE PRICES AND
DISCOUNTS

Buying items on sale is a great way to save money.
In this lesson you will calculate your savings by purchasing a skateboard at a percent-off sale. Then you will use discount coupons to shop for sporting equipment.

Determine Savings

How can you calculate your savings in a percent-off sale?

You want to join the after-school skateboard club this year, but you need to purchase the required equipment in order to participate. Skates on Seventh is advertising a percent-off sale. You have $175 to spend. Refer to this advertisement in answering the following questions.

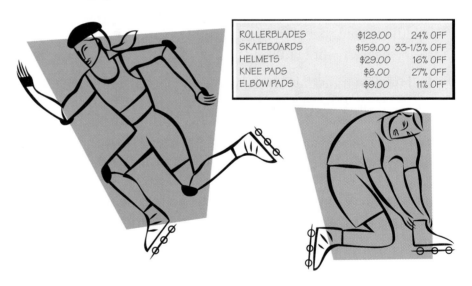

ROLLERBLADES	$129.00	24% OFF
SKATEBOARDS	$159.00	33-1/3% OFF
HELMETS	$29.00	16% OFF
KNEE PADS	$8.00	27% OFF
ELBOW PADS	$9.00	11% OFF

1 Make a quick estimate to see if you have enough money to buy a skateboard, helmet, knee pads, and elbow pads.

2 Use your calculator to figure out how much you will save for each item.

3 Figure out the sale price of each item.

Shop with Discount Coupons

The sports club wants to buy a variety of sports equipment for students to try out. They want you to buy as many new pieces of equipment as you can for their budget of $350. Fortunately, they have lots of discount coupons for you to use.

You need to buy *at least* three different kinds of equipment and use a different coupon from the handout Discount Coupons for each one. Remember, you can use each coupon only once and you can't use more than one coupon per type of equipment. The goal is to spend close to $350 without going over.

1 Decide which items you want to buy and which coupons you will use for each.

2 Make a table to show each original cost, the discount from the coupon, and the sale price.

3 Figure out the total cost of your purchases.

4 Figure out how much you saved by buying the items with discount coupons.

BASKETBALL	$23.00
CHAMPIONSHIP BASKETBALL	$52.00
VOLLEYBALL	$44.00
VOLLEYBALL NET	$109.00
CATCHER'S MITT	$51.50
CATCHER'S MASK	$15.70
BASEBALLS	$39.50 PER DOZEN
BASEBALL BAT	$26.95
BATTING HELMET	$11.85
SOCCER BALL	$26.30
TENNIS RACKET	$49.55
TENNIS BALLS	$12.00 FOR 3 CANS
FOOTBALL	$18.00
HOCKEY STICK	$49.00
HOCKEY PUCK	14.00 FOR 2

> **How much money will you save by shopping with discount coupons?**

Design a Sale Advertisement

Create a colorful flyer announcing a sale at your favorite department store. For each advertised item, include:

- the original price
- the percent off, discount, and sale price

hot **words** | discount price

HW**omework**

page 44

12 Percent Smorgasbord

A budget is a useful tool for keeping track of your money. In this lesson you will be given a set amount of money to plan an athletic banquet. You will need to plan a menu, select entertainment, and purchase awards, decorations, and gifts for the coaches. Finally, you will use all of your data to display your budget in a circle graph.

Use a Budget to Plan an Athletic Banquet

How can you work with a budget to make planning decisions?

You have been given $2,500 to plan the athletic banquet. The money was donated for the banquet and that's all it can be used for, so you need to spend close to $2,500. You can't spend more than this.

Athletic Banquet Attendance

Sport	Students	Coaches
Field hockey	20	2
Football	50	3
Soccer	30	2
Tennis	16	2
Basketball	20	2
Volleyball	12	1
Cross-country	12	1
Swimming	8	1
Lacrosse	18	2

1. Use the handout Athletic Banquet Price List to help you plan your choices for food, entertainment, awards, decorations, and gifts for the coaches.

2. When you have figured out how much you will spend for each category, write your plan for what you will be doing for each of the following: food, entertainment, awards, decorations, and gifts for the coaches.

3. Make a chart to record your total expenses in each category.

4. Write down the total amount you will spend on the banquet. Tell how much money, if any, you will have left over from the $2,500 you were given to spend.

Display Your Budget Data in a Circle Graph

You need to present the information in your banquet plan to the athletic advisory committee. Make a circle graph for the presentation.

- Use the circle on the handout Circle Graph.

- Use percents to label each sector of your circle graph.

- Make sure your sectors are clearly labeled and that the graph has a title.

How can you make a budget presentation?

Write an Article for *Buyer Beware*

Write an article for *Buyer Beware* magazine called "A Banquet on a Budget."

- Your article should include your chart and circle graph.

- Include tips to readers on how to make choices that save money.

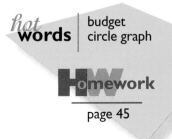

hot **words** | budget
circle graph

HW**omework**

page 45

What's the Best Buy?

Applying Skills

Use your calculator to find the unit price for each of the following. Round to the nearest cent.

1. 7 oz of crackers for $1.19

2. 14 oz of cottage cheese for $1.19

3. 16 boxes of raisins for $5.60

Find the better buy based on unit price.

4. A 35-oz can of Best Brand Plum Tomatoes is on sale for $0.69. A 4-lb can of Sun Ripe Plum Tomatoes is $1.88.

5. A can of Favorite Dog Food holds 14 oz. Four cans are $1.00. The price of three cans of Delight Beef Dog Food, each containing 12 oz, is $0.58.

6. For each item, predict which is the better buy. Then use paper and pencil or a calculator to find the better buy.

	Item	Jefferson Auto Stores	Tom's Auto Parts
a.	oil	12 qt for $10.99	6 qt for $5.99
b.	antifreeze	12 oz for $3.79	6 oz for $1.79
c.	auto wax	6 cans for $14.29	5 cans for $12.98

Extending Concepts

7. Six cans of fruit drink are on sale for $1.95. Individually, the price of each can is $0.35. How much does Tanya save buying 6 cans on sale?

8. Tubes of oil paint can be bought in sets of 5 for $13.75 or bought separately for the unit price. What would be the price of 2 tubes of this oil paint?

9. The price of three bottles of Bright Shine Window Cleaner, each containing 15 oz, is $2.75. Two bottles of Sparkle Window Cleaner, each containing 18 oz, can be purchased for $1.98. Which is the better buy?

Writing

10. Create an advertisement for orange juice in which a small-size carton on sale is a better buy than a larger-size carton at regular price.

The Best Snack Bar Bargain

Applying Skills

The price graph below shows the unit prices for three different shampoos. Aloe Shampoo is $1.00 for 2 oz, Squeaky Clean is $2.75 for 3.5 oz, and Shine So Soft is $3.75 for 4 oz.

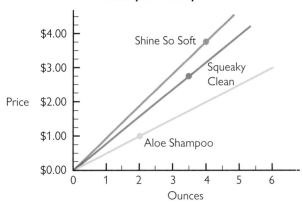

Shampoo Price per Ounce

1. Use the graph to find the price of 4 oz of Aloe Shampoo.

2. Use the graph to find the price of 0.5 oz of Shine So Soft.

3. Use the graph to find out which shampoo has the lowest unit price.

4. Use the graph to find out which shampoo has the highest unit price.

5. Use your calculator to find the price of 1 oz of each shampoo. Use the graph to check your calculations.

Extending Concepts

Use the data for the products listed below to construct a price graph.

Product	Number of Units	Price
Crunchy Crackers	9.5 oz	$1.20
Buzzy Tuna	5.5 oz	$1.00
Bessy's Pancake Mix	12 oz	$1.60
Pino's Imported Pasta	8 oz	$2.00
Lean ground beef	9 oz	$2.40

Use the price graph you constructed to answer the following questions.

6. What is the price of 5 oz of Crunchy Crackers?

7. What is the price of 1 oz of Buzzy Tuna?

8. What is the price of 2 oz of Bessy's Pancake Mix?

9. What is the price of 14 oz of Pino's Imported Pasta?

10. What is the price of 15 oz of Lean Ground Beef?

Writing

The information below is missing data that is needed to complete a problem. Tell what might be missing. Make up data that could be used to complete and solve the problem.

11. The price of a box of biscuits is $0.89. On the box it says, "New larger size— 15 ounces."

Cheaper By the Pound

Applying Skills

Find the price per pound to decide the better buy.

Potatoes	
2 lbs	$1.09
5 lbs	$2.69
10 lbs	$3.59
20 lbs	$6.29

1. What is the price per pound of each package of potatoes?

2. Which size package of potatoes is the best buy in terms of price per pound?

Calculate the price per pound of the items below.

	Item	Price per Item	Weight in Pounds	Price per Pound
3.	Bike	$179.99	45	
4.	Rollerblades	$135.99	11	
5.	Basketball	$ 24.99	2	
6.	1996 complete set of baseball cards	$ 26.99	3.6	
7.	Earrings	$ 16.00	0.25	
8.	Watch	$ 34.95	0.25	
9.	Pearl ring	$ 79.99	0.13	

10. For which of the items above would you pay the least per pound?

Extending Concepts

11. Alejandro bought an 18-lb watermelon for $4.00. To the nearest cent, what is the price per pound?

12. A 5-lb bag of dog food sells for $3.85. Maurice's dog eats 2 bags of dog food every month. What is the monthly price per pound of the dog food?

Making Connections

13. Select several magazines or newspapers. Find out how much a subscription costs to each of the magazines or newspapers. Compare the unit price to the newsstand price.

It Really Adds Up

Applying Skills

Crunchy Popcorn	No-Ad Popcorn
$0.45 per bag	$0.35 per bag

Find the price of buying one bag of popcorn every day for a week.

1. Crunchy Popcorn

2. No-Ad Popcorn

Find the price of buying one bag of popcorn every day for a month (4.3 weeks).

3. Crunchy Popcorn

4. No-Ad Popcorn

Find the price of buying one bag of popcorn every day for a year.

5. Crunchy Popcorn

6. No-Ad Popcorn

7. How much would you save if you bought No-Ad Popcorn instead of Crunchy Popcorn for:

a. 1 week

b. 1 month (4.3 weeks)

c. 1 year

8. If you bought a bag of No-Ad Popcorn every day instead of Crunchy Popcorn, how many days would it take to save $30? $75? Explain how you figured it out.

Extending Concepts

9. At Jacy's Market, you can get 5 mangos for $1.95. At Nia's Market, you can get 3 for $1.29. Are the mangos cheaper at Jacy's or at Nia's?

10. You can get 3 cans of Mei Mei's Soup for $1.23 and 2 cans of Pei's Soup for $0.84. Which brand costs less per can?

11. Why would a store owner price an item at $9.99 for 5 instead of $2.00 each?

Making Connections

12. Select two supermarket advertisements from a newspaper. Compare the prices of similar items. Which store seems to have the better buys? Give reasons for your answer.

Quilting Ratios

Applying Skills

Write each ratio as a fraction in lowest terms.

1. 6 to 8   **2.** 8:44

3. $\frac{60}{32}$ **4.** 20 to 30

Write two ratios that are equal to each ratio.

5. 5:30   **6.** 12:15

7. 30:12   **8.** 8:2

Using the quilt below, write each ratio as a fraction in lowest terms. Then tell whether it is a part-to-part or part-to-whole ratio.

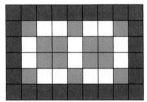

9. light blue squares to dark blue squares

10. dark blue squares to all squares

11. light blue squares to all squares

Extending Concepts

The data below show how some students spent their time from 4 P.M. to 5 P.M. yesterday. Decide if statements **12–15** are true or false.

How Students Spent Their Time

	Number of Students
Homework	ꝢꝢꝢꝢꝢꝢ ꝢꝢꝢꝢꝢ
Sports practice	ꝢꝢꝢꝢꝢ
Music practice	ꝢꝢꝢꝢꝢ
Chores or job	ꝢꝢꝢꝢꝢ I
Other	IIII

12. One out of every three students did homework.

13. One out of every five students did chores.

14. The ratio of students doing homework to students practicing music is 5 to 2.

15. The ratio of students doing chores to students practicing music or sports is 2 to 3.

Making Connections

16. Sports statistics use ratios to describe a player's performance. Choose several players in a sport you enjoy. Research the players' statistics and write several ratios for each set of statistics. For example, baseball ratios might include hits to at bats or total bases to hits. Football ratios could include field goals made to field goals attempted.

Compare the ratios for players you have chosen. Do the ratios explain why one player is more valuable than another?

In the Mix

Applying Skills

Use any method to solve each problem.

1. This pattern shows one "octave" on a piano keyboard.

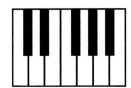

a. Electronic keyboards come in different sizes. One keyboard has 4 octaves. How many black keys does it have?

b. Amy's favorite keyboard has 35 white keys. How many black keys does it have?

2. The scale on a map is 3 inches = 4 miles.

a. How far is the actual distance between two towns that are 10.5 inches apart on the map?

b. How many inches apart are two streets that actually are one mile apart?

Use ratio tables to solve each problem.

3. Anya's class is selling wrapping paper. For every 5 rolls they sell, they make a profit of $1.80. She wants to figure out how many rolls the class needs to sell to make a profit of $270. Fill in the missing numbers in her ratio table.

Number of rolls	5	10	15	30	d.
Profit	$1.80	a.	b.	c.	$270

4. A passenger jet travels at an average speed of 450 miles per hour. Use the ratio table to find the missing times or distances.

Time (hours)	1	2	4	c.	10
Distance (miles)	a.	b.	1,800	2,700	d.

Extending Concepts

5. The price of first-class postage is 37¢ for up to one ounce, 54¢ for up to two ounces, 71¢ for up to three ounces, 88¢ for up to four ounces, and $1.05 for up to five ounces.

Marco claims the table below is a ratio table. Is he correct? Explain.

Weight (ounces)	1	2	3	4	5
Cost	37¢	54¢	71¢	88¢	$1.05

Making Connections

6. In question 5, the number of ounces increases by one, and the cost increases 17¢. Even though there is an adding pattern in both sets of numbers, the numbers do not stay in the same ratio to each other.

Describe another real-life example that has a constant adding pattern but does not stay in the same ratio to each other.

Halftime Refreshments

Applying Skills

Determine whether each pair of ratios is proportional. Write *yes* or *no*.

1. $\frac{3}{4}, \frac{9}{16}$

2. $\frac{4}{6}, \frac{6}{9}$

3. $\frac{20}{16}, \frac{15}{12}$

4. $\frac{7}{12}, \frac{8}{15}$

Solve each proportion.

5. $\frac{4}{6} = \frac{n}{21}$

6. $\frac{12}{n} = \frac{8}{15}$

7. $\frac{n}{28} = \frac{30}{14}$

8. $\frac{4}{n} = \frac{n}{16}$

Write a proportion and solve each problem.

9. Merchants price their products based on proportions. Suppose 12 cans of soda cost $4.80. What is the price of 36 cans of the same soda?

10. A tree casts a shadow that is 28 feet long. Liza notices that at the same time, her shadow is 3 feet long. She knows that she is $5\frac{1}{2}$ feet tall. If the ratio of the height of the tree to the length of its shadow is proportional to the ratio of Liza's height to the length of her shadow, how tall is the tree?

11. There are onions and green peppers in a bag of frozen mixed vegetables. The ratio of ounces of onions to ounces of green peppers is 4 to 9. How many ounces of onions are there if there are 30 ounces of green peppers?

12. Suppose you buy 2 CDs for $21.99. How many CDs can you buy for $65.97? Assume all of the CDs cost the same amount.

Extending Concepts

13. How many proportions can you make using only the numbers 1, 3, 4 and 12?

14. At Friday's football game, the athletic boosters sold three times as many hot dogs as brownies. Altogether, 500 hot dogs and brownies were sold. How many of each item were sold?

15. Challenge Which plot of land is most square? Explain your answer.

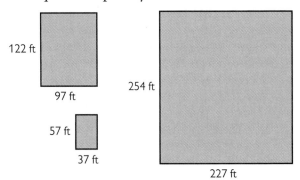

Making Connections

16. Explore exchange rates in currency (money) from different countries.

a. Choose three different countries. Find out the current exchange rate between the currency used in that country and American dollars.

b. For each country, set up a proportion to calculate how much money from that country you would get for $100 American dollars.

Can I Use a Proportion?

Applying Skills

Determine whether the following sets of numbers are proportional to each other. Write *yes* or *no*. Explain your reasoning.

1.

1st number	2	4	6	8	10
2nd number	4	6	8	10	12

2.

1st number	3	6	9	12	15
2nd number	4	8	12	16	20

3.

1st number	2	3	5	8	12
2nd number	1	1.5	2.5	4	6

4.

1st number	0	3	4	8	12
2nd number	2	5	6	10	14

5. For each of the problems above, make a line graph of the table of values. You should end up with four separate lines.

6. Two of the problems above have numbers that are proportional. What is true about the line graphs of these two problems?

7. Two of the problems above have numbers that are not proportional. How are the line graphs of these two problems different from the line graphs of the proportional sets of numbers?

Extending Concepts

8. Jack is 7 years younger than his sister Nancy. This year on their birthdays, he turned 7 and she turned 14. Nancy noticed that she was twice as old as Jack. She wondered, "Will I ever be twice as old as Jack again? If so, when?"

a. Make a table of their ages for the next five years.

b. For each year, calculate the ratio of Nancy's age to Jack's age as a single number rounded to the nearest hundredth. What do you notice about the ratios?

c. When will Nancy be twice as old as Jack again? Explain your answer.

d. Is there a time when Nancy will be one and a half times as old as Jack? If so, when? If not, explain why not.

e. Will the ratio of their ages ever become 1? Explain why or why not.

Writing

9. Robert wants to make hot cocoa for his friends. He knows he needs 2 spoonfuls of cocoa mix for each mug of hot cocoa that he makes. Robert thinks: "I'll make sure the number of spoonfuls is always one more than the number of mugs I want to make." Do you agree with Robert's reasoning? Explain why or why not.

Team Spirit

The field hockey team's total expenses for last year were $10,000. Estimate how much was spent on each of the expenses listed below. Use benchmarks and mental math to make your estimates.

	Item	Percent of Budget	Estimated Cost
1.	Uniforms	22%	
2.	Transportation	4%	
3.	Coach's Salary	12%	
4.	Equipment	49%	
5.	Officials' Fees	6%	
6.	Trainer's Salary	7%	

Estimate each number in items **7–10.** Then use your calculator to see how close your estimate is.

7. 49% of 179

8. 24% of 319

9. 19% of 354

10. 34% of 175

11. Find 7% of $400.

7% means_____ for every_____

12. Find 12% of $300

12% means_____ for every_____

13. To estimate 24% of 43, LeRon substituted numbers and found 25% of 44. His answer was 11. Using his calculator, he found that the exact answer is 10.32. LeRon concluded that substituting numbers causes you to overestimate. Do you agree? If not, give a counterexample.

14. Nirupa calls home from college at least once a week. A 30-minute phone call costs $10.00 on weekdays. Nirupa can save 20% if she calls on a weekend. How much money does she save on a 30-minute call made on Saturday?

15. Look through newspapers and magazines to find articles involving percents. Design a collage with the articles. Write out percents from 1 through 100 and their equivalent fractional benchmarks.

Playing Around

Applying Skills

The circle graph below shows the budget for the middle school production of *The Music Man.*

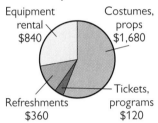

Budget for
Once Upon a Mattress

Equipment rental $840

Costumes, props $1,680

Refreshments $360

Tickets, programs $120

Use the circle graph to estimate what percent of the total cost was spent on each of the following:

1. costumes and props

2. tickets and programs

3. refreshments

4. equipment rental

Use your calculator to help you find the actual percent of the total cost that was spent on each of the following:

5. costumes and props

6. tickets and programs

7. refreshments

8. equipment rental

Use the following information to make a circle graph.

9. Ticket sales for *The Music Man* totaled $560. Students collected the following amounts: Vanessa $168, Kimiko $140, Ying $112, Felicia $84, and Norma $56. Label the circle graph, using names and the percents collected. Give the graph a title.

Extending Concepts

Step-in-Time shoe store took in the following amounts in January:

Men's dress shoes	$750
Women's dress shoes	$1,500
Children's sneakers	$2,000
Adult athletic shoes	$850

The circle graph below was made using the information above.

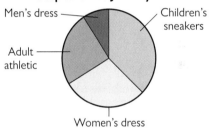

Step-in-Time January Sales

Men's dress

Children's sneakers

Adult athletic

Women's dress

10. Was the circle graph made correctly? Explain.

Making Connections

Percentage of the World's Population by Continent

Africa	12%
Europe	14%
North America	8%
Asia	60%
South America	6%

Source: *World Almanac 1996*

11. Use the handout Circle Graph to make a circle graph of the world's population.

Sale Daze

Applying Skills

Use a calculator to find the discount and sale price for items **1–4**. Round to the nearest cent.

1. regular price: $78
rate of discount: 25%

2. regular price: $29
rate of discount: 20%

3. regular price: $45
rate of discount: 15%

4. regular price: $120
rate of discount: 37%

Which shop has the better buy?

	Item	Super Sports	Sammy's Sport Shop
5.	Football	Regular price: $39.95 rate of discount: 10%	Regular price: $42.95 rate of discount: 15%
6.	Basketball	Regular price: $36.50 rate of discount: 20%	Regular price: $35 rate of discount: 15%
7.	Helmet	Regular price: $19.95 rate of discount: 20%	Regular price: $17.95 rate of discount: 15%

8. Orit needed a helmet to skate on the half-pipe at Skateboard Plaza. She bought one for 60% off the regular price of $31.50. How much did she save? How much did she pay?

9. Salvador paid $27.50 for a catcher's mask that was on sale. The regular price was $36.90. What was the discount?

Extending Concepts

10. Kai-Ju has saved $40 to buy a tennis racket that regularly sells for $59.99. She read an ad that announced a 25% discount on the racket. Has she saved enough money to buy it? If so, how much will she have left over? If not, how much more does she need?

11. Ryan went into a hardware store that had a 20%-off sale. His two purchases originally had prices of $38.75 and $7.90. How much did Ryan save because of the sale?

Writing

Make up the missing data in items **12** and **13**. Then write a word problem that can be solved using the data.

12. roller skates
regular price: $89.99
rate of discount:

13. tennis racket
regular price: $45.89
rate of discount:

Percent Smorgasbord

Applying Skills

For items **1–4**, how many items on the list can the shopper buy without overspending?

1. shopper has $140
discount: 12%

video game set	$89.95
video game cartridge	$29.50
blow dryer	$27.50
sweater	$34.00

2. shopper has $150
discount: 10%

roller skates	$99.00
compact disk	$12.99
jacket	$75.00
shirt	$18.00

3. shopper has $180
discount: 25%

bicycle	$129.00
bike helmet	$ 39.00
tennis racket	$ 89.00
ring	$ 58.00

4. shopper has $62
discount: 50%

radio	$29.00
jeans	$32.00
earrings	$18.00
shoes	$44.00

5. A box of Munchy Cereal contains 24 oz of cereal. It is on sale for 50% off the regular price of $4.80. Toasty Cereal, which contains 50% more cereal than Munchy Cereal, is $4.80 per box.

Which box has the lower price per ounce? To the nearest cent, how much less is the price per ounce for this box?

Extending Concepts

Use your calculator to find the total for each bill. The tip is 15%.

6. **South of the Border**

Beef tacos	$6.25
Guacamole	$4.50
Chicken tamale	$7.50

7. **Thai Cuisine**

Chicken with ginger	$7.95
Sweet and sour chicken	$8.90
Shrimp and baby corn	$9.95

Writing

8. Answer the letter to Dr. Math.

Dear Dr. Math,

Neely's Hot Dogs advertises that their hot dogs contain more protein than fat. Their hot dogs contain proteins, carbohydrates, and fat. Each hot dog contains 11 g of fat and 2 g of carbohydrates. This makes up 55% of the hot dog's content. Could the advertisement be true? How can I tell?

Lois Kallory

ALLPLAY

To: Apprentice Game Designers
From: AllPlay Company Directors

Welcome to AllPlay. As an apprentice game designer, part of your role will be to test games for us. We have lots of exciting assignments in store for you. You will be trying out new games, analyzing them, improving them, and creating your own. To help you design games that are fair and fun to play, you will be learning about probability and statistics.

Good luck!

PHASE**ONE**

Games of Chance and Probability

In this phase, you will explore the meaning of chance. You will play different booths at a carnival using coins and cubes, and investigate one of the booths in depth. Using strip graphs, frequency graphs, and probability lines, you will learn to show chance visually.

What mathematics is involved in testing and analyzing games of chance?

CHANCE ENCOUNTERS

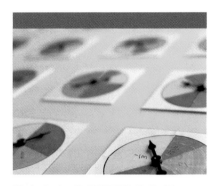

PHASE**TWO**
Spinner Games and Probability

Games played with circular spinners are the focus of this phase. You will experiment with ways of changing games to improve your chances of scoring. To solve the Mystery Spinner Game, you will create spinners to match a set of clues. The clues describe probabilities in words and numbers. Finally you will be ready to design your own Mystery Spinner Game.

PHASE**THREE**
Fair and Unfair Games

How can you tell if a game is fair? To test your predictions about how fair a game is, you will conduct experiments. You will learn to use outcome grids to determine the probability of outcomes in a game. Your understanding of fractions, decimals, and percentages will help you to compare the probabilities of scoring in games.

PHASE**FOUR**
Real-World Simulations

Your final project is to design, test, and present your own simulation game. But first, you will test and design other simulation games. In a Miniature Golf Simulation, you will compare your results to actual data from that game. In the Shape Toss Game, you will collect data by tossing a penny. Then you will create a simulation to match your game results.

PHASE ONE

To: Apprentice Game Designers
From: AllPlay Company Directors

Your first assignment is to work on six new booths in the Carnival Collection. We have designed a rough draft, and now it is time to test it.

It's hard to tell about a booth from just one person's experience. That's why we want all of you to try the booths and tell us what you think. We want your ideas on how to improve them.

ALLPLAY

In this phase, you will test games of chance and use probability to analyze the chances of winning.

To help you in your role as game tester and designer, you can think about how likely it is that certain events will happen every day. Do you think what happens to you is due to chance or luck?

Games of Chance and Probability

WHAT'S THE MATH?

Investigations in this section focus on:

THEORETICAL and EXPERIMENTAL PROBABILITY

- Understanding the concept of chance
- Conducting probability experiments
- Determining the theoretical probability and experimental probability of events

MULTIPLE REPRESENTATIONS of PROBABILITY

- Representing data with frequency and strip graphs
- Describing probabilities with qualitative terms, such as *likely, unlikely,* and *always*
- Describing probabilities quantitatively with decimals, fractions, and percentages

MODELING SITUATIONS with SIMULATIONS

- Ranking event probabilities on a scale of 0 to 1

MathScape Online

mathscape2.com/self_check_quiz

1 The Carnival Collection

The AllPlay Company is developing games of chance designed as booths in the Carnival Collection. You will test the Carnival Collection and graph the class results. Can you come up with ideas for improving the Carnival Collection?

Test the Carnival Collection

What can you find out about games by comparing the experiences of many players?

Play the Carnival Collection with a partner. Each player should make a score sheet. For each turn, record the name of the booth and whether or not you scored a point.

The Carnival Collection

Players take turns. On your turn, pick the booth you want to play. You can choose a different booth or stay at the same booth on each turn. The player with the most points at the end of ten turns wins.

- Get Ahead Booth: Toss a coin. Heads scores 1 point.

- Lucky 3s Booth: Roll a number cube. A roll of 3 scores 1 point.

- Evens or Odds Booth: Roll a number cube. A roll of 2, 4, or 6 scores 1 point.

- Pick a Number Booth: Predict what number you will roll. Then roll a number cube. A true prediction scores 1 point.

- Coin and Cube Booth: Toss a coin and roll a number cube. Tails and 3, 4, 5, or 6 scores 1 point.

- Teens Only Booth: Roll a number cube 2 times. Make a 2-digit number with the digits in any order. A roll of 13, 14, 15, or 16 scores 1 point.

Design a Winning Strategy

Use the graph of your class results to help you design a strategy that would give players a good chance of scoring points in the Carnival Collection.

Which booths give you the best chances of scoring points?

1 Which booths would you go to? Why?

2 How many points do you think a player would be likely to score in 10 turns by using your strategy? Explain your thinking.

3 How many points do you think a player would be likely to score in 100 turns by using your strategy? Show how you know.

Improve the Carnival Collection

Use what you have learned to improve the Carnival Collection.

1 Pick one booth at which you think it is too hard to score points. Change the booth to make it *more likely* for players to score points.

2 Pick one booth at which you think it is too easy to score points. Change the booth to make it *less likely* for players to score points.

3 Design a new booth that uses coins and/or number cubes. What do you think the chances are of scoring points at the booth? Explain your thinking.

hot **words** | chance outcome

Homework

page 80

2 Coins and Cubes Experiment

Would everyone get the same results in 20 turns at the same booth in the Carnival Collection? To find out, your class will choose one booth and conduct an experiment on it. You will make predictions, gather data, and graph the data.

Conduct an Experiment

What can an experiment tell you about your chances of scoring in a game?

Write down your predictions about the booth your class has chosen to test. How many points do you think you will get in 20 turns at the booth? What do you think the greatest number of points in the class will be? the least? What do you think the most common number of points will be? After making your predictions, conduct an experiment.

1 With a partner, take turns playing the booth. Before each toss of a coin or cube, predict what you will get.

2 After each turn, record the results on a strip graph as shown. Color the box for that turn if you scored a point. Leave the box blank if you did not score. Play for 20 turns.

H	T	H	H	H	T	T	H	T	T	H	H	H	H	H	T	H	H	H	T	T	T

3 Record the total number of points and greatest number of points scored in a row.

Summarize the Results

Use the results from your class's strip graphs and frequency graph to figure out the class totals for points, no points, and tosses. Then answer the following questions:

1 Use the whole class's results to find the experimental probability for getting a point. Then find the experimental probability for your individual results. How do your results compare with the class results?

2 What is the theoretical probability of getting a point at these booths: Get Ahead, Lucky 3s, Evens or Odds, and Pick a Number? Explain how you figured it out.

3 What do the results of the class experiment tell you about the chances of winning points at the booth?

What Is Probability?

Probability describes the chances of an event occurring. For example, the *theoretical probability* of getting heads when you toss a coin is:

$$\frac{\text{Number of favorable outcomes}}{\text{Total number of possible outcomes}} \longrightarrow \frac{\text{heads}}{\text{heads and tails}} \longrightarrow \frac{1}{2}$$

It could also be written as 1 out of 2, 50%, or 0.5.

To figure out *experimental probability,* you need to collect data by doing an experiment. One class tossed a coin 400 times with these results:

Number of heads: 188	Number of tails: 212	Number of tosses: 400

Based on that class's results, the experimental probability of getting heads is:

$$\frac{\text{Total number of times the favorable outcome happened}}{\text{Total number of times you did the experiment}} \longrightarrow \frac{\text{heads was tossed}}{\text{coin tosses}} \longrightarrow \frac{188}{400}$$

It could also be written as 188 out of 400, 47%, or 0.47.

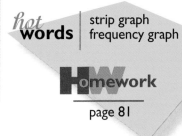

hot **words** | strip graph
frequency graph

Homework
page 81

3 From Never to Always

DESCRIBING THE
LIKELIHOOD
OF EVENTS

How likely are different events on a typical school day? In this lesson, you will use words and numbers to describe and compare probabilities for events. Then you will analyze the results of experiments for a new Carnival Collection booth.

Use Numbers to Represent Probabilities

How can you use percentages, fractions, and decimals to describe probabilities?

Imagine that the name of each student in your class is in a hat. The teacher will pick one name from the hat without looking.

1 Use a percentage, fraction, or decimal to describe the probability of each event:

  a. The winner is a girl.

  b. The winner is a boy.

  c. You are the winner.

  d. The winner is a student in your class.

2 Make a probability line and place the events on it.

Probability Lines

- Probability lines are useful tools for ordering events from *least likely* to *most likely* to happen.

- Events that are *impossible* and will *never* happen have a probability of 0, or 0%.

- Events that are *definite* and will *always* happen have a probability of 1, or 100%.

- A probability can never be greater than 1, or 100%.

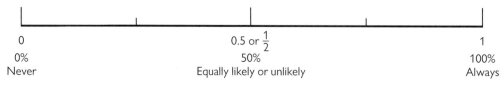

0	0.5 or $\frac{1}{2}$	1
0%	50%	100%
Never	Equally likely or unlikely	Always

Analyze a New Carnival Collection Booth

The handout High/Low Booth describes a new booth two students have designed and tested for the Carnival Collection. Use the information on the handout to write a report on the booth.

How could you use what you have learned about probability to analyze a new booth?

1 Write a summary of the class experiment that answers the following questions:

   a. What was the most common number of wins? the range of wins?

   b. How do Dan and SooKim's results compare with those of their classmates?

   c. What is the experimental probability of scoring a point, based on Dan and SooKim's data? Explain how you figured it out.

   d. What is the experimental probability of scoring a point, based on the whole class's data?

2 Figure out the theoretical probability of scoring a point.

   a. Explain how you figured out the theoretical probability.

   b. Why is the theoretical probability different from the experimental probability that you figured out?

3 Make a probability line to show the theoretical probabilities of winning a point at these three booths: High/Low, Evens or Odds, and Lucky 3s. See page 50 for the rules for Evens or Odds and Lucky 3s. Label each probability with the booth name and with a fraction, decimal, or percentage.

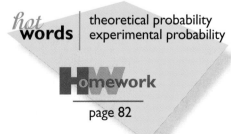

hot **words** | theoretical probability
            | experimental probability

Homework

page 82

PHASE TWO

You might imagine yourself as a detective in this phase as you create spinners by solving sets of clues. This prepares you to design your own Mystery Spinner Game at the end of the phase. Exchanging the game you create with your classmates and playing games adds to the fun.

As you look at spinners, you will see that they show events of unequal probability. What are some ways to change a game to make the probability more equal?

Spinner Games and Probability

WHAT'S THE MATH?

Investigations in this section focus on:

THEORETICAL and EXPERIMENTAL PROBABILITY

- Conducting probability experiments
- Determining the theoretical probability of events
- Distinguishing events that have equal probabilities from events that have unequal probabilities

MULTIPLE REPRESENTATIONS of PROBABILITY

- Exploring area models of probability
- Describing probabilities with qualitative terms, such as *likely, unlikely,* and *always*
- Describing probabilities quantitatively with decimals, fractions, and percentages
- Relating verbal, visual, and numerical representations of probability

MathScape Online
mathscape2.com/self_check_quiz

4 Spin with the Cover-Up Game

INVESTIGATING
CHANCE USING
A SPINNER

What is it like to play a game with circular spinners that have unequal parts? You will find out when you compare your results in the Cover-Up Game with those of your classmates. How can you improve the Cover-Up Game?

Design a New Game Card

How can you improve the chances of finishing the game in fewer spins?

After playing the Cover-Up Game, think about the class results. Use the results to help you design a new game card that will give you a good chance of finishing in the fewest spins.

- Draw a game card with 12 empty boxes.

- In each box write *red*, *blue*, or *yellow*. You do not need to use every color.

The Cover-Up Game

1. Play with a partner using a spinner like the one shown. You will also need to make a Game Card and an Extras Table like the ones shown.

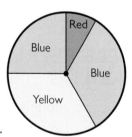

2. Take turns spinning a color. On your Game Card, cover a box of that color with an X. If all boxes of that color are covered, mark a tally in the Extras Table.

3. The game ends when every box on your Game Card is covered with an X. Record the number of spins you took. The goal is to make an X in all the boxes with as few spins as possible.

Game Card

B	B	B	B
R	R	R	R
Y	Y	Y	Y

Extras Table

B	R	Y

Analyze Spinners

For each spinner shown, what are the chances of landing on each part of the spinner? Use words and numbers to describe the chances.

How can you use what you know about spinners to design a new game?

Foods

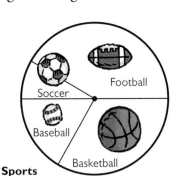

Sports

Games

How would you describe a spinner using words and numbers?

Design a New Cover-Up Game

Design your own version of the Cover-Up Game and write about what happened when you played it.

1 Design a circular spinner with 3–6 parts.

   a. Label the parts with the names of music groups, places, or whatever you'd like.

   b. Show the size of each spinner part by labeling it with a fraction.

2 Make a game card for your spinner that gives you a good chance of making an X in all of the boxes in the fewest spins.

3 Play your game and record the extras you get in a table. Record the number of spins.

4 How did you design your game? Write about why you filled in the game card the way you did.

Tip: To make a spinner, trace the circle and center point from the Cover-Up Game spinner. Then divide up the parts the way you want.

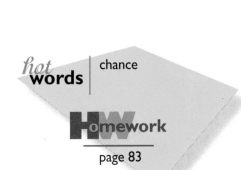

hot **words** | chance

Homework

page 83

5 The Mystery Spinner Game

How could you create a spinner using a set of clues?
This is what you will do as you play the Mystery Spinner Game.
Writing about your strategies and solving problems with clue sets
will help you design your own Mystery Spinner Game later on.

How can you design a spinner to match probabilities described in words and numbers?

Play the Mystery Spinner Game

After you have solved Clue Set A with the class, you will be given
Mystery Spinner Game Clue Sets. Solve the Mystery Spinner
Game Clue Sets in your group. Be ready to explain the strategies
you used to figure out clues and design spinners.

A You have the same chances of getting a stuffed animal as getting a T-shirt.

A You are likely to win an apple about 50% of the time.

A The spinner has three kinds of prizes.

A You will probably get a T-shirt about $\frac{1}{4}$ of the time.

The Mystery Spinner Game

1. Each player gets one clue. Players read the clues
   aloud to the group. They cannot show their clues
   to one another.

2. The group draws one spinner that matches all
   the players' clues.

3. The group labels the parts of the spinner with
   fractions, decimals, or percentages.

4. The group checks to make sure the spinner
   matches all the clues.

Interpret Clues and Describe Strategies

Draw a circular spinner that matches Clue Set B. Label the parts with fractions, percentages, or decimals. Then answer these questions:

- What strategies did you use to make one spinner that matched all the clues?

- What kinds of clues make good starting points?

- How can you prove that your spinner matches all of the clues?

What strategy can you use to design a spinner that matches a set of clues?

B This spinner has four kinds of sports equipment that you can win. The chances of getting a basketball are $\frac{1}{6}$.

B You will probably get skates about 40 times in 240 spins.

B The chances of getting a Frisbee™ are about 0.33.

B You are twice as likely to get sneakers as a basketball.

Fix Sets of Clues

Explain what's wrong with Clue Sets C and D. Change the clues to fix the problems. Then draw a spinner to match each set of corrected clues.

How can you fix clues that are misleading?

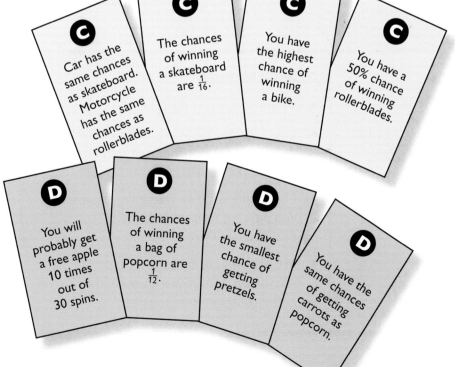

C Car has the same chances as skateboard. Motorcycle has the same chances as rollerblades.

C The chances of winning a skateboard are $\frac{1}{16}$.

C You have the highest chance of winning a bike.

C You have a 50% chance of winning rollerblades.

D You will probably get a free apple 10 times out of 30 spins.

D The chances of winning a bag of popcorn are $\frac{1}{12}$.

D You have the smallest chance of getting pretzels.

D You have the same chances of getting carrots as popcorn.

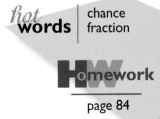

hot **words** | chance fraction

Homework
page 84

6 Designing Mystery Spinner Games

The AllPlay Company would like you to design some new Mystery Spinner Games. You have learned a lot about spinners and probability. Here is a chance for you to use what you know to create games that are fun and challenging.

Design a Mystery Spinner Game

Can you write a set of clues that another student could use to draw a complete spinner?

As a class you will discuss the Unfinished Mystery Spinner. After you have completed the Unfinished Mystery Spinner Game, use the Mystery Spinner Game guidelines to design your own game. When you write your clues, use a separate sheet of paper. Keep the spinner hidden from your classmates.

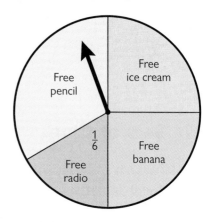

Unfinished Mystery Spinner

Mystery Spinner Game

Spinner Guidelines

- Draw a spinner with 3–5 parts.

- Label the spinner parts with the names of foods, music groups, sports, or whatever you'd like.

- Show the size of each spinner part by labeling it with a fraction, percentage, or decimal.

Clue Guidelines

- Write at least four clues.

- Use a variety of words and numbers to describe the probability of spinning each spinner part.

- Make sure you use fractions, decimals, and percentages to describe probabilities in your clues.

- Make sure your clues tell your classmates everything they need to know to draw your spinner.

Test a Mystery Spinner Game

Try out your partner's Mystery Spinner Game and write your feedback. The goal is to help each other improve the games.

How can you test a game designed by someone else?

1 Exchange clues with a partner. (Keep the spinner hidden.)

2 Draw a spinner to match your partner's clues.

3 Give your partner feedback on the game by writing answers to these questions:

   a. What did you like about the game?

   b. What made the game easy or difficult to solve?

   c. Were the clues missing any information? If so, what information would you like to get from the clues?

   d. What suggestions would you give to improve the game?

Tips for Giving Feedback

1. Remember, you are giving feedback to help someone else do a better job or to figure something out.

2. Put yourself in your partner's shoes. What kind of feedback would you find helpful?

Revise the Mystery Spinner Game

Read your partner's feedback on your game. Think about ways to solve any problems.

1 Revise your game. You may need to change your clues and/or your spinner. If your partner drew a spinner that is different from the one you drew, you may want to change your game so that there is only one solution.

2 Write about your game. Include answers to the following questions.

   a. How did you improve your game? What changes did you make? Why?

   b. What mathematics did you use in your game?

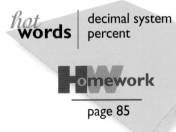

hot **words** | decimal system
               | percent

Homework

page 85

PHASE THREE

The games in this phase are more complex. The results depend on combining the outcomes of two events.

Suppose you have one spinner divided into thirds, with the numbers 1–3, and another spinner divided into fourths, with the numbers 1–4. In a game, Player A gets a point when the sum of two spinner values is odd and Player B gets a point if the sum is even. How might an outcome grid be used to show all the possible combinations in the game?

Fair and Unfair Games

WHAT'S THE MATH?

Investigations in this section focus on:

THEORETICAL and EXPERIMENTAL PROBABILITY

- Conducting probability experiments
- Comparing experimental and theoretical probabilities
- Determining theoretical probability by showing all possible combinations on outcome grids
- Applying probability to determining fairness

MULTIPLE REPRESENTATIONS of PROBABILITY

- Using visual models to analyze theoretical probabilities
- Exploring numerical representations of probabilities and connecting them with visual models

MathScape Online
mathscape2.com/self_check_quiz

7 Is This Game Fair or Unfair?

EVALUATING THE FAIRNESS OF A GAME

Imagine you are playing a game of chance and you keep losing. You begin to think that the game is unfair, but your opponent insists that it is fair. In this lesson, you will use mathematics to figure out if a game of chance is fair.

Use Outcome Grids to Determine Chances

How can you determine each player's chances of getting points?

Make an outcome grid for the Special Sums Game to show all the possible sums when you roll two number cubes. Follow these directions to make your grid.

1. Draw an outcome grid like the one shown and fill in the boxes. Use the grid to figure out the probability of getting each of the sums (2–12). Which sum do you have the best chance of getting?

2. Color or code the grid to show the ways each player can get points. Use a different color or symbol for each player.

3. When you have finished the grid, write about each player's probability of getting points.

Special Sums Game

1. Players take turns rolling two number cubes and adding the two numbers together.

   If the sum is 1, 2, 3, or 4, Player A gets 1 point. If the sum is 5, 6, 7, or 8, Player B gets 1 point. If the sum is 9, 10, 11, or 12, Player C gets 1 point. Players can get points on another player's turn. For example, if any player rolls a sum of 10, Player C gets 1 point.

2. Record the number of points each player gets.

3. Each player gets 5 turns. The player with the most points at the end of 15 turns is the winner.

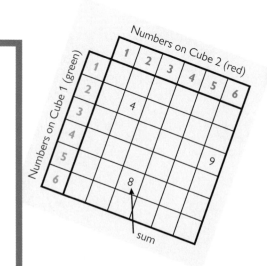

Change the Special Sums Game

If the game is unfair, change the rules to make it fair. If it is fair, then make it unfair. To change the game, follow these steps.

1 Draw a new outcome grid and fill in all the possible sums in the Special Sums Game.

2 How do you want to change the rules? Use the grid to help you figure out which sums get points for which players. Complete the rules below.

> • Players take turns rolling two number cubes. Add the two numbers.
> • Player A gets 1 point when the sums are _____.
> • Player B gets 1 point when the sums are _____.
> • Player C gets 1 point when the sums are _____.
> • The player with the most points after 15 rolls wins. (Each player gets 5 turns.)

3 Color or code the grid to show the ways each player can get a point.

When you have finished changing the game, explain why your new game is fair or unfair.

hot **words** | outcome grid

Homework
page 86

8 Charting the Chances

Sometimes games that look fair are really unfair.
Sometimes the opposite is true. How can you determine which games are fair? In this lesson, you will collect data and use outcome grids to analyze spinner, coin, and number-cube games.

Investigate the Fairness of Different Games

How can you determine whether a variety of games are fair?

You will be given several different games to play with a partner or in groups of three. Your challenge is to figure out whether these games are fair or unfair. For each game, follow these steps:

1 Make predictions. Before you play the game, predict whether it is fair or unfair. If you think the game is unfair, which player do you think has the advantage? Write about your thinking.

2 Collect data. Play the game and record your results.

3 Make an outcome grid for the game. Color or code the grid to show each player's probability of getting a point.

4 Describe each player's probability of getting a point in one of these ways: fraction, decimal, or percentage.

5 Write your conclusions. Explain your thinking about whether the game is fair or unfair.

Hint: Some of the games use two game pieces. Others use one, such as a spinner that players spin twice. For games with one piece, you can use the top of the outcome grid for the first spin or toss and the side of the outcome grid for the second spin or toss.

Make an Unfair Game Fair

No one wants to play the Sneaky Sums Game because it is unfair. The AllPlay Company would like you to make the game fair. Read the handout Sneaky Sums Game.

How could you change the rules to make the game fair?

1 On Centimeter Grid Paper, make an outcome grid to find out how unfair the game is. What is each player's probability of getting a point?

2 Change the rules to make the game fair. Which sums do you want to score points for which players? Write your new rules.

3 Color or code the grid to show how each player can get points with your new rules. Describe each player's probability of getting points in two of these ways: fractions, decimals, or percentages.

When you are finished, describe how you changed the rules to make the game fair.

Write About Outcome Grids

How would you teach someone else about outcome grids? Write a letter to a student who will have this unit next year. Share what you have learned about outcome grids. Use one of the games you played in this lesson as an example. Make sure to include answers to these questions in your letter:

- How can you figure out how to set up an outcome grid for a game? How would you label the top and side of the grid? How can you figure out how many boxes the grid should have?

- How can you use an outcome grid to figure out a player's probability of getting points?

- How can you use an outcome grid to figure out if a game is fair?

- What are some mistakes that people might make when they are setting up an outcome grid for a game? How can you avoid those mistakes?

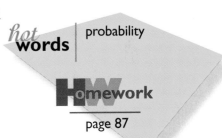

hot **words** | probability

Homework
page 87

Which Game Would You Play?

COMPARING THE CHANCES OF WINNING

Which of four unfair games do you have the best chance of winning? To decide, you will use fractions, decimals, and percentages to describe and compare probabilities. Then you will analyze the Carnival Collection booths from Lesson 1.

Rank Probability Grids for Unfair Games

How can you compare chances of scoring in games with different numbers of outcomes?

You can picture the probability of winning a game by shading a grid. This grid shows that the probability of winning this game is $\frac{6}{36}$ or $\frac{1}{6}$.

$\frac{1}{6}$

1 Look at Grids for Unfair Games. Describe Terry's probability of getting a point in each game in at least two different ways: fractions, decimals, or percentages.

2 Rank the grids from "Best for Terry" to "Worst for Terry" (best is a 1). How did you figure out how to rank the grids?

3 For each game, how many points do you think Terry would be likely to get in 100 turns? How many points would Kim be likely to get? Explain your thinking.

4 Make a grid that is better for Terry than the second-best game, but not as good as the best game.

Grids for Unfair Games

Here are grids for four different unfair games. If you were Terry, which game would you want to play?

A   B   C   D

☐ Kim gets a point

▨ Terry gets a point

Compare Carnival Collection Probabilities

Read over the rules for the Coin and Cube Booth and Teens Only Booth in the Carnival Collection on page 50. Then complete the following:

Read over the rules for the Coin and Cube Booth and Teens Only Booth in the Carnival Collection on page 50.

1 What is the probability of scoring a point at the Coin and Cube Booth? at the Teens Only Booth?

   a. Make an outcome grid for each booth. Color or code the grid to show a player's probability of getting a point.

   b. Describe the theoretical probability of scoring a point in three ways: fractions, decimals, and percentages.

2 How many points do you think you would get at the Coin and Cube Booth and the Teens Only Booth for the following number of turns? Explain how you made your predictions.

   a. 36 turns

   b. 72 turns

   c. 100 turns

   d. 1,800 turns

3 Make a probability line to show the theoretical probability of scoring a point at each of the six booths. Rank the booths from greatest to least chances of scoring a point (best is 1).

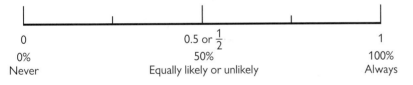

0	0.5 or $\frac{1}{2}$	1
0%	50%	100%
Never	Equally likely or unlikely	Always

4 At each booth, if a player does *not* score a point, it means that the booth owner wins. Change the rules of the Teens Only Booth so that it is almost equally fair for the player and the booth owner. Write your new rules and explain why your version is more fair.

How can you use what you have learned to compare probabilities?

hot **words** | theoretical probability

HW**omework**
_____
page 88

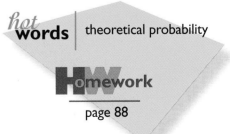

PHASE **FOUR**

Many games are designed to simulate things that people really do. For example, there are computer games that simulate flying a plane or planning a city. Each of these games captures, in a simple way, some of the events in the actual activity.

Think about a familiar activity. How could you design a simulation that is like the real-world activity?

Real-World Simulations

WHAT'S THE MATH?

Investigations in this section focus on:

THEORETICAL and EXPERIMENTAL PROBABILITY

- Conducting probability experiments
- Collecting and analyzing data to determine the probabilities of real-world events
- Using grid models to represent and analyze theoretical probabilities

MULTIPLE REPRESENTATIONS of PROBABILITY

- Representing probability verbally, visually, and numerically

MODELING SITUATIONS with SIMULATIONS

- Modeling the probabilities of real-world events by creating simulations
- Testing simulations by comparing the probabilities in a simulation and its real-world counterpart
- Applying probability and statistics to making simulations more realistic

MathScape Online
mathscape2.com/self_check_quiz

10 Is the Simulation Realistic?

The AllPlay Company has designed a new Miniature Golf Simulation game. A simulation is a game that is like a real activity. The *events* in a simulation represent what would happen in the real activity. The likelihood of events in the simulation should be the same as their likelihood in the real activity.

Play a Simulation Game

How can you collect data on how often each event occurs in the simulation?

Play the Miniature Golf Simulation with your group. The group will need the handouts Miniature Golf Simulation and Miniature Golf Score Sheet and two different-colored number cubes.

Miniature Golf Simulation

- Players take turns. On your turn, roll the number cubes and read the game grid to find out how many strokes it took to get the ball in the hole.

- Record the number of strokes on your score sheet.

- The player with the least number of strokes after 18 holes wins.

How to Read the Game Grid

If you roll a 3 on the red cube and a 2 on the green cube, you will get a "Wimpy Putt" which means it took you 3 strokes to get the ball in the hole. In miniature golf, low scores are better than high scores.

Compare Simulation Data to Actual Data

Refer to the handout Actual Miniature Golf Games to complete the following items.

1 Look at the scores on the frequency graphs your class made after playing the simulation. How do the class scores from the simulation compare with the scores on the graph on the handout for the actual game?

2 Complete the table by figuring out the Average Times per Game and the Experimental Probability for each event.

$$\text{average times per game} = \frac{\text{total occurrences}}{\text{number of games}}$$

$$\text{experimental probability for each event} = \frac{\text{total times each event happened}}{\text{total number of holes played}}$$

3 Compare the experimental probabilities of the actual game to the theoretical probabilities of events in the simulation.

a. Use the data from the table you completed in item **2** to rank the events from most likely to least likely.

b. Figure out the theoretical probabilities of events using the Miniature Golf Simulation. Rank the theoretical probabilities of events from most likely to least likely to occur.

4 How realistic is the simulation? Use data to support your conclusions.

How can you compare the probabilities in the simulation to the actual sport?

Revise the Simulation

Use the data from the actual miniature golf games to design a more realistic simulation.

- Make a new grid for the simulation. How many boxes do you want to give each event?

- What is the theoretical probability of each event in your new simulation?

- Write a description telling how you decided what changes to make.

Is it as likely to get a hole-in-one in the simulation as the actual game?

hot **words** | frequency graph
statistics

Homework

page 89

11 The Shape Toss Game

What are the steps involved in designing a simulation? To find out, you will play a tossing game. You will analyze data your class gathers. Figuring out how many boxes on a grid to give each event helps make the simulation realistic.

Collect Data on a Tossing Game

How can you collect class data for a tossing game?

Play the Shape Toss Game. To record your results, make a table like the one shown.

How can you compile and analyze the class data?

The Shape Toss Game

1. Each group needs the Shape Toss Game Board and a penny for each player to toss.

2. Each player gets at least 5 turns to toss the penny onto the board.

3. Players score the following points if the penny lands entirely inside one of the shapes on the board:

   - Lands in a triangle = 20 points
   - Lands in a rectangle = 10 points
   - Lands in a hexagon = 5 points
   - Misses = 0 points

4. Players record their results.

Sample Table

Player	Triangle (20 pts.)	Rectangle (10 pts.)	Hexagon (5 pts.)	Miss (0 pts.)	Total Score
Corey	0	2	1	2	25 pts
Delia	1	2	2	1	50 pts
Miguel	2	0	2	2	50 pts
Totals	3	4	5	5	125 pts

Design a Simulation of the Tossing Game

In the Shape Toss Game, there are four possible events: land in a triangle, land in a rectangle, land in a hexagon, and miss. Follow these steps to analyze the class data and figure out the probability of each event.

How can you use class data to design a realistic simulation of the tossing game?

1 Copy and complete a table like the one shown so that it has your class's data.

Table of Our Class's Data for the Shape Toss Game			
Number of students who played: _____			
Total number of tosses: _____ (Each student got 5 tosses.)			
Average score: _____			
Event	**Total Times the Event Happened**	**Average Times the Event Happened per Student**	**Experimental Probability of Event**
Triangle			
Rectangle			
Hexagon			
Miss			

2 Make a probability line to rank the events from least likely to most likely.

3 Draw a blank, 36-box grid for your simulation of the Shape Toss Game. Use the class data to help you figure out how many boxes to give each event. Remember that the probabilities of events in the simulation should match as closely as possible the probabilities from actually playing the game.

hot **words** | simulation
experimental probability

Homework
_____
page 90

12 Real-World Simulation Game

Now it's your turn to design your own simulation. This final project brings together all that you have learned about probability and statistics in this unit. It's exciting to create, test, revise, present, and reflect about your own simulation game.

Brainstorm Ideas

How can you design a simulation of a real-world activity?

Design a simulation game. When you have completed all the steps, use the Project Cover Sheet to summarize your project.

1 Plan your simulation game. Choose a real-world activity to simulate. Describe your ideas and how you will collect data using the Project Planning Sheet.

2 Collect data on the real-world activity. Display the data you collected and describe how you collected it. Write a summary of what you found out.

3 Make a probability line. Use your data to rank the likelihood of the real-world events from least likely to most likely.

4 Design a grid game board for your simulation. Decide which two game pieces (cubes, spinners, etc.) you will use to play your simulation. Draw a blank grid with the appropriate number of boxes. Fill in your grid to make a realistic simulation. Color or code the events.

5 Describe the probabilities of events in your simulation. Use fractions, percentages, or decimals.

6 Write the rules. What is the goal? How do players score points? When does the game end? You may want to design a score sheet.

7 Try out your simulation game. Play it at least twice and record the results. Describe what happened when you played. This isn't an accurate test of the simulation, but it will give you a sense of what it's like to play.

Test the Simulations

Trade games with a partner and play your partner's simulation to test it. Write answers to these questions to give your partner feedback.

- What do you like about the simulation?

- What suggestions would you like to give the game designer?

- How realistic is the simulation?

Giving Constructive Feedback

Here are some ways to give feedback that will help your classmates improve their simulations.

"I really like how you did...."

"Some things you could improve are...."

"I had trouble understanding what you meant by...."

"The part that seemed unclear to me was...."

How did your partner make the simulation realistic?

Reflect on the Projects

Think about the experiences you had when you designed your simulation game and your partner tested it. Consider these questions as you write about your experiences.

- What did you find out when other students tested your simulation?

- What do you like best about your simulation?

- How would you change your simulation to make it more realistic?

- How would you change your simulation to make it more fun to play?

- What tips would you give to other students who wanted to make a realistic simulation game?

hot **words** | natural variability
law of large numbers

Homework

page 91

The Carnival Collection

Applying Skills

In the Get Ahead Booth, you score a point if your coin comes up heads.

1. Predict what your score would be if you played 40 turns at Get Ahead. Explain your prediction.

2. Predict what your score would be if you played 500 turns at Get Ahead. Explain your prediction.

3. Use a coin to play 40 turns at Get Ahead. How many points did you score? Compare your actual results to your predictions for 40 turns.

4. The Tally Sheet shows the points scored in 10 turns at several booths. Which booth would you choose to go to? Why?

Tally Sheet

Lucky Spins	IIII
Big Four	II
Toss 'n' Roll	I
No Doubles	ЖНⅠ III

Extending Concepts

At a new booth called Five or More, you roll a number cube and score a point if you get a 5 or 6.

5. How many points do you think a player would be likely to get in 30 turns at the Five or More Booth? Explain your reasoning.

6. Two students each played 30 turns at the Five or More Booth and recorded their scores in this table. How does each player's score compare with your prediction?

Student	Number of Points Scored
Raphael	8
Sabitra	13

7. How could you change the rules of Five or More so that players would be more likely to want to play it?

8. Design a booth for the Carnival Collection that uses a number cube or coin. Which gives the better chance of scoring points, your booth or Five or More? Why do you think so?

Making Connections

9. Nim is a game for two players that originated in China thousands of years ago. In one variation of Nim, 15 match sticks are laid in a row.

Each player in turn removes 1, 2, or 3 match sticks. The player who is forced to take the last match stick is the loser. Do you think that this game involves chance, skill, or both? Explain your thinking.

Coins and Cubes Experiment

Applying Skills

Find the experimental probability of scoring a point. Express your answers as fractions.

1. A person at the Get Ahead Booth scores 31 points in 50 turns.

2. A person at the Evens and Odds Booth scores 27 points in 72 turns.

3. A person at the Lucky 3s Booth scores 10 points in 49 turns.

4. The strip graph shows what happened when Elena played 10 turns at Lucky 3s. A point is shown by a shaded square.

   a. How many times did she score?

   b. What is her experimental probability of scoring a point?

   ☐☐☐☐☐☐☐☐☐☐

5. Twelve people each played 10 turns at the Get Ahead Booth. The frequency graph shows how many points each player got.

   ```
                  X        (X: one person's results)
             X  X
          X  X  X  X
          X  X  X  X     X
       ─────────────────────────
       1  2  3  4  5  6  7  8  9  10
               Number of Wins
   ```

   a. What was the lowest number of points?

   b. What was the highest number of points?

   c. What was the most common number of points?

Extending Concepts

Five people each played 10 turns at a carnival booth and recorded their data on these strip graphs. A point is shown by a shaded square.

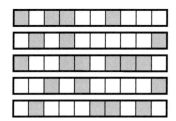

6. Make a frequency graph to show how many points each person got.

7. What is the most common number of points? the average number of points?

8. Combine the data from all five strip graphs to figure out the experimental probability of getting a point.

Writing

9. Answer the letter to Dr. Math.

   Dear Dr. Math,
   I don't get it! My friend and I each flipped a coin 10 times. My friend got 8 heads and I got only 2 heads. We thought that we each had a 50% chance of getting heads, but neither of us got 5 heads! Why did this happen? Should we practice flipping coins?
   Sooo Confused

From Never to Always

Applying Skills

These four names of students are put in a box: Mica, Syed, Anna, and Jane. One name is picked from the box, and that person wins a prize. Find the theoretical probability that the winner's first name:

1. starts with a vowel

2. contains an *e*

3. contains a vowel

4. has more than four letters

5. does not end with an *e*

6. Order the theoretical probabilities of the five events above from least likely to most likely by placing them on a probability line. Write the probability as a fraction or percentage.

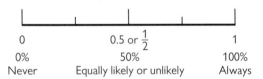

In the Get Ahead Booth, you score a point if your coin comes up heads. In the Lucky 3s Booth, you score a point if you roll a 3 on a number cube. Here are the results for 12 turns at each booth.

Get Ahead: T, T, H, T, H, H, T, H, T, H, H, H

Lucky 3s: 1, 4, 2, 5, 3, 6, 1, 4, 3, 5, 1, 2

7. Find the experimental probability of scoring a point at each booth. Give each answer as a fraction, a decimal, and a percentage.

Extending Concepts

8. Read the ad and answer the following questions.

   a. Do you think this is a good deal? Why or why not?

   b. What do you think your chances of winning would be?

Making Connections

9. A **tetrahedron** is a geometric figure with four identical sides. An **octahedron** has eight identical sides. Suppose you have a tetrahedron die whose sides are numbered 1 to 4 and an octahedron die whose sides are numbered 1 to 8. What is the probability of rolling a 1 on each die? Explain your thinking.

Tetrahedron   Octahedron

Spin with the Cover-Up Game

Applying Skills

Draw a circular spinner for each pair of conditions. Label each part with a fraction.

1. Green and yellow are the only colors. You have an equal chance of spinning green or yellow.

2. You will spin yellow about half of the time. You have an equal chance of spinning either of the other colors, red or blue.

3. You are twice as likely to spin green as yellow. You will spin yellow about $\frac{1}{3}$ of the time.

4. Which spinner shown below gives you the best chance of landing on "Striped"? Use a fraction to describe the probability.

5. Which spinner shown below gives you the highest probability of landing on "Dotted"? Use a fraction to describe the probability.

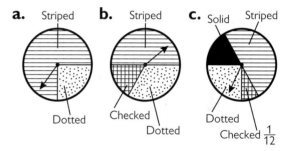

Extending Concepts

Suppose you are playing the Cover-Up Game using the spinner shown. Each time the spinner lands on a color, players make an X in the box of that color on their game cards.

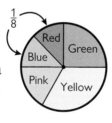

6. Find the probability of getting each color on the spinner.

7. Make a 24-box game card that would give you a good chance of finishing the game in the fewest possible spins. Explain your thinking.

8. Make a 24-box game card that would give you a poor chance of finishing the game in the fewest possible spins. (The card should still allow you to finish the game.)

Writing

9. Answer the letter to Dr. Math.

Dear Dr. Math,

I can't seem to win at the Cover-Up Game. When I make my game card, I put the colors or words that I think will come up the most in the boxes near the center of the card. I put the ones that won't come up much in the corners. This strategy doesn't seem to work well. Can you suggest a better one?

Wants T. Winn

The Mystery Spinner Game

Applying Skills

Find the probability of green, red, and yellow for each spinner.

1.
   Red
   Green

2.
   Yellow
   Red Green

3.
   Yellow
   Red Green

Draw a spinner for each set of parts and label the parts. If the parts cannot be used to make a spinner, explain why not.

4. $\frac{1}{2}, \frac{1}{4}, \frac{1}{8}, \frac{1}{8}$ 5. $\frac{1}{3}, \frac{1}{4}, \frac{1}{4}, \frac{1}{6}$ 6. $\frac{1}{3}, \frac{1}{12}, \frac{3}{4}$

7. Design a spinner to match the following clues. Label the parts with fractions.

   - There are four pizza toppings.

   - You will get pepperoni about 25% of the time.

   - You will get pineapple about 30 times in 180 spins.

   - You are twice as likely to get onion as pineapple.

   - You have the same chances of getting mushroom as pepperoni.

Extending Concepts

8. At Jamila's Unusual Ice Cream Shop, customers can spin a spinner to try to win a free cone. If they do not win a free cone, they must buy the flavor they land on. Draw a spinner that matches these clues. Label the parts with fractions.

   - There are three possible flavors.

   - Free cone is half as likely as walnut.

   - The chance of getting kiwi is $\frac{1}{4}$.

   - You will get pumpkin about 15 out of 40 times.

9. What's wrong with this set of clues? Describe what is wrong and change the clues. Draw the spinner to match your corrected clue set.

   - There are four trips you could win.

   - You will win a trip to New Zealand about 7 times out of 20 spins.

   - You have a 20% chance of winning a trip to Hawaii.

   - You are twice as likely to win a trip to Paris as a trip to Hawaii.

   - You are more likely to win a trip to Alaska than a trip to Hawaii.

Writing

10. Answer the letter to Dr. Math.

> Dear Dr. Math,
> My friend and I are arguing about the Cover-Up Game. She says she can make a new spinner and a game card with 12 boxes so that she will always cover the card in exactly 12 spins. That doesn't seem possible to me, but she insists she can. How can she do it?
> Ida N. Know

Designing Mystery Spinner Games

Applying Skills

Write at least three clues to describe each spinner. Use different ways to describe the probabilities—fractions, decimals, or percentages.

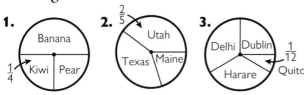

1.
2. $\frac{2}{5}$
3.

Extending Concepts

David designed the spinner shown and wrote the following clues to describe it.

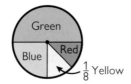

- There are four colors.

- You have the highest chance of getting green.

- The chances of getting yellow are the same as the chances of getting red.

- You are twice as likely to get blue as red.

David's clues do not give enough information for someone else to draw exactly the same spinner.

4. Draw a spinner that matches David's clues, but is different from his spinner. Label each part with a fraction.

5. Write a new set of clues that gives all the information needed for someone else to draw David's spinner.

Making Connections

The table shows the names and sizes of four deserts in different parts of the world. Sizes are given to the nearest 10,000 square miles.

Desert	Location	Size (mi²)
Arabian	Egypt	70,000
Chihuahuan	Mexico, United States	140,000
Sonoran	Mexico, United States	70,000
Taklimakan	China	140,000

6. Draw a spinner with one part for each desert in the table. Let the size of each spinner part correspond to the size of the desert. For example, the spinner part for the Chihuahuan Desert should be twice as big as the part for the Arabian Desert.

7. Show the size of each part by labeling it with a fraction or percentage.

8. Write at least four clues to describe your spinner. Use words and numbers to describe the probability of spinning each part. Make sure your clues give all the information needed to draw your spinner.

Is This Game Fair or Unfair?

Applying Skills

In the Spinner Sums Game, players spin the two spinners shown and add the two numbers. Player A gets a point if the sum is a one-digit number. Player B gets a point if the sum is a two-digit number. The player with the most points after 15 turns wins.

Spinner A

| 5 | 7 |
| 8 | 10 |

Spinner B

| 2 | 5 |

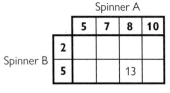

	Spinner A			
	5	**7**	**8**	**10**
2				
5			13	

Spinner B (row labels)

1. Make an outcome grid like the one shown. Fill in all the sums.

2. List all the different sums you can get. What is the probability of getting each of the sums when you spin the two spinners?

3. Color or code the grid to show the way each player can get points.

4. What is each player's probability of getting a point?

5. Is the game fair or unfair? Explain why.

Extending Concepts

Imagine that you did an experiment where you tossed two number cubes 1,800 times and recorded the number of times you got each sum. Use an **outcome** grid that shows all the possible sums when you roll two number cubes to help you answer these questions.

6. Which sum or sums would you get the most? the least? Why?

7. Which of these bar graphs do you think your results would look the most like? Why?

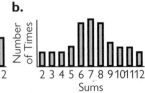

a.

b.

c.

8. How many 9s do you think you would get? Why?

9. How many 3s do you think you would get? Why?

Making Connections

Each number in the *Fibonacci Sequence* is the sum of the previous two:
1, 1, 2, 3, 5, 8, …

A prime number has no factors other than 1 and itself: 2, 3, 5, 7, …

In Roll Again, players roll two number cubes. Player A gets a point if the sum is in the Fibonacci Sequence. Player B gets a point if the sum is a prime number, but not in the Fibonacci Sequence. Player C gets a point otherwise. The player with the most points after 20 turns wins.

10. Make an outcome grid for this game.

11. Color or code the grid to show the way each player scores points.

12. Who do you think will win? Why?

Charting the Chances

Applying Skills

1. An outcome grid for two coin flips is shown below. Use the grid to find the probability that you will get:

a. two heads

b. one head and one tail

c. two tails

d. the same result on both coins

Penny

	H	**T**
H	HH	HT
T	TH	TT

Dime

2. Draw an outcome grid for a game in which you flip a coin and roll a number cube. Use the grid to find the probability that you get:

a. heads and a 3

b. tails and an even number

Extending Concepts

In the Spin and Roll game, players roll a number cube and spin the spinner shown. They use the two numbers that show up to make the *smallest* two-digit number they can. For example, a player who rolls a 4 and spins a 3 will make the number 34.

Player A gets a point if the number is less than 20. Player B gets a point if the number is more than 40. Player C gets a point if the number is between 20 and 40.

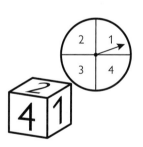

3. Make an outcome grid for Spin and Roll.

4. Color or code the grid to show how each player scores points.

5. Describe each player's probability of getting points as a fraction, decimal, and percentage.

6. Change the rules to make this game fair for three players. Explain your thinking.

7. Write a set of rules that makes this a fair game for four players.

Writing

8. Answer the letter to Dr. Math.

Dear Dr. Math,

I designed a number-cube game called Double or Nothing. Players roll two cubes. They win a prize if they get doubles. I need to figure out how many prizes to buy. I expect that the game will be played about 200 times. About how many prizes do you think I will give out? Tell me how you figured it out, so that I can do it myself next time.

Booth Owner

Which Game Would You Play?

Applying Skills

These numbers represent probabilities of winning different games.

$\frac{11}{25}$	2 out of 5	84%	0.12

1. For each probability, make a 5-by-5 grid on Centimeter Grid Paper. Add shading to the appropriate numbers of boxes to show each probability.

2. Rank the grids from the greatest chance of winning to the least chance of winning.

3. What is the probability of *not* winning for each game?

These outcome grids represent four games. Shaded squares mean Player A gets a point; unshaded squares mean Player B gets a point.

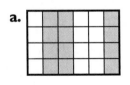
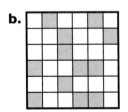
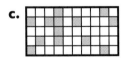
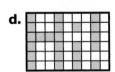

4. Describe Player A's probability of getting a point in each game as a fraction, a decimal, and a percentage.

5. Rank the grids from "Best for A" to "Worst for A."

Extending Concepts

At the Flip and Roll Booth, players flip a coin and roll a number cube. A player scores a point if the coin comes up heads and the cube shows a 1 or 2.

6. Make an outcome grid for Flip and Roll.

7. Shade the grid to show a player's probability of getting a point.

8. How many points do you think you would get at Flip and Roll in 60 turns? in 280 turns? in 2,500 turns? How did you make your predictions?

9. If a player does not get a point, this means that the booth owner wins. Write a new set of rules for Flip and Roll so that it is equally fair for the player and the booth owner. Explain why your version is fair.

Making Connections

The Native American game called *Totolospi* was played by the Moki Indians of New Mexico. Players used flat throwing sticks which were plain on one side and painted on the other. Players would throw three sticks. They scored points if all sticks fell with the same side up.

10. Is it possible to use an outcome grid to find the probability of all sticks falling the same way? Why or why not? If not, can you think of any other way to solve this problem?

Is the Simulation Realistic?

Applying Skills

Two students collected scorecards from a miniature golf course that allows players to take up to six strokes to get the ball in the hole. They used the data to design a new miniature golf simulation. Here is their grid:

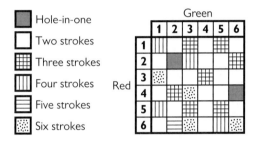

Hole-in-one
Two strokes
Three strokes
Four strokes
Five strokes
Six strokes

1. Maylyn rolled a red 5 and a green 2. Which event did she get?

2. What is the theoretical probability of each event in the simulation? Explain how you figured it out.

3. Make a probability line to rank the events in the simulation from least likely to most likely.

4. There are five possible events in the simulation: out, single, double, triple, and home run. Find the theoretical probability of each event.

5. Make a probability line to order the events in the simulation from least likely to most likely.

Compare the simulation to actual data for a professional baseball team.

6. In 6,000 times at bat, the team got 4,200 outs. What is their experimental probability of getting an out? Explain how you figured it out.

7. What is the professional team's experimental probability of getting a home run? In 6,000 times at bat, the team got 180 home runs.

8. How many boxes would you give to outs on the 36-box grid to make the simulation more realistic? How many boxes would you give to home runs? Explain how you found your answers.

Extending Concepts

A simulation baseball game is shown in the outcome grid. The game involves rolling two number cubes.

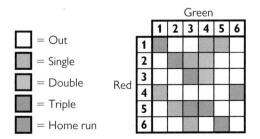

= Out
= Single
= Double
= Triple
= Home run

Writing

9. Define *simulation* in your own words.

The Shape Toss Game

Applying Skills

In a simulation, how many boxes would you give to an event on a 40-box outcome grid if its experimental probability is:

1. 25%   **2.** 0.75   **3.** 65%   **4.** 0.48

In the Shape Toss Game, players toss a penny onto a board and score points if the penny lands inside a shape. Here is the data from one class, where the total number of tosses was 90.

Event	Total Times Event Happened in 90 Tosses
Triangle	10
Rectangle	17
Hexagon	20
Miss	43

5. Find the experimental probability of each event.

6. How many boxes would you give each event on a 50-box simulation grid?

Extending Concepts

In the Shape Toss Game shown, players score 20 points for triangles, 10 points for rectangles, 5 points for hexagons, and 0 points for misses.

7. Use your experimental probability data to find the total number of points scored in 90 tosses. How did you figure it out?

8. What was the average number of points scored per toss?

9. Design a simulation of the Shape Toss Game based on the experimental probability of each event.

a. Choose two game pieces to play the game, and make an outcome grid with the appropriate number of boxes.

b. Color or code the grid to give a realistic simulation. Explain how you decided how many boxes to give each event.

c. Play your simulation twice and record the results. Describe what happened.

Writing

10. Answer the letter to Dr. Math.

Dear Dr. Math,

Maria and Carla play on my basketball team. Our coach had to pick a player to take a foul shot. I thought he would pick Maria because she has made 100% of her foul shots this season. But he picked Carla, who has made only 60% of her shots this season. Can you explain this? I thought 100% was better than 60%. Here are their statistics:

| Maria | 2 shots attempted | 2 shots made |
| Carla | 100 shots attempted | 60 shots made |

Team Manager and Record Keeper

Real-World Simulation Game

Applying Skills

1. For each pair of game pieces below, give the number of boxes in their outcome grid. Then make the grid.

   a. two number cubes

   b. a number cube and a coin

2. Find the experimental probability of each event in the table. Tell how many boxes you would give to each event in a 6 × 6 outcome grid.

Serves in Ten Games of Tennis

Event	Number of Times
Fault	12
Double fault	6
No fault	24
Ace	6

Extending Concepts

The data in the table refer to the NCAA soccer championships played from 1959 to 1990. The events are the number of goals scored in the final by the winning team.

Number of Goals Scored by Winning Team in Final	Total Goals Scored in 31 Games
1	10
2	10
3	4
4	5
5	2

3. Find the experimental probability of each event. Make a probability line and rank the events from least likely to most likely.

4. Choose two game pieces you will use to play the game. Make an outcome grid with the appropriate number of boxes. Fill in your grid to make a realistic simulation. Explain how you decided how many boxes to give to each event.

5. Write a short set of rules explaining how to use your simulation.

6. What are the probabilities of each event happening in your simulation? Use fractions, decimals, and percentages to describe these probabilities.

Writing

7. Answer the letter to Dr. Math.

> Dear Dr. Math,
>
> My friend and I want to design a simulation softball game. We want the scores in the simulation to be like the scores real middle school students get. How can we collect the data we need? How much data should we collect?
>
> Hi Score

What is involved in writing a mathematical argument?

MAKING
MATHEMATICAL
ARGUMENTS

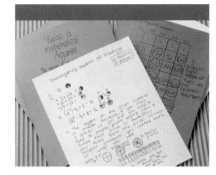

PHASE**ONE**
Signs, Statements, and
Counterexamples

When a statement about
numbers is always true, we call
it a rule. One way to show that a
statement is true is to model it
with objects. In this phase, you
will use cubes to help you think
about statements about adding,
subtracting, multiplying, and
dividing with signed numbers
and answer the question: Is it
always true?

PHASE**TWO**
Roots, Rules, and Arguments

Another way to show that a
statement is always true is
to look for special cases. In
this phase, you will examine
different mathematical
arguments about squares,
cubes, and roots to look for
special cases (0, 1, proper
fractions, and negative
numbers). This prepares you to
write your own mathematical
arguments.

PHASE**THREE**
Primes, Patterns, and
Generalizations

Patterns can be used as a basis
for a rule and to explain how
you know a rule is true beyond
just giving examples. By the
end of this phase, you will
choose an interesting pattern
(or possible pattern) and
experiment with it. These
patterns involve squares, cubes,
primes, factors, and multiples.

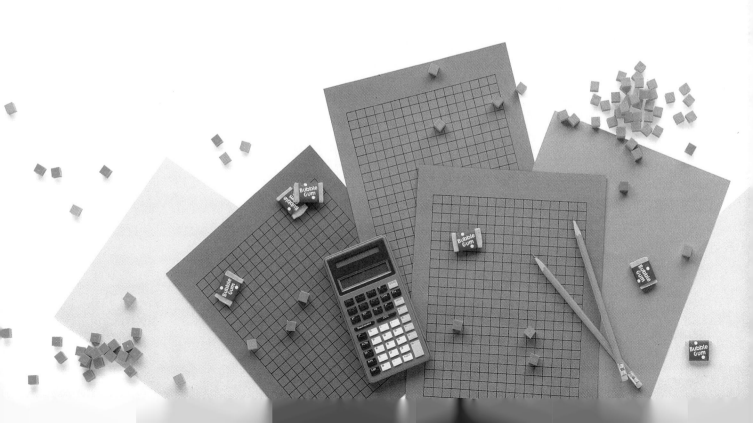

PHASE ONE

A counterexample shows that a statement is not always true. For example, a counterexample to the statement "All apples are red" is the statement "Some apples are green." Why might knowing how to come up with a counterexample be useful?

Understanding and using signed numbers is an important skill. In what professions might you use this skill?

Signs, Statements, and Counterexamples

WHAT'S THE MATH?

Investigations in this section focus on:

NUMBER and OPERATIONS

- Understanding signed numbers

- Understanding how to add, subtract, multiply, and divide integers

MATHEMATICAL REASONING

- Making and evaluating mathematical statements about positive and negative number operations

- Looking for counterexamples to determine whether a mathematical statement is untrue

NUMBER SYSTEMS

- Understanding how addition, subtraction, multiplication, and division are related to one another

MathScape Online
mathscape2.com/self_check_quiz

1 Statements About Signs

EXPLORING INTEGER STATEMENTS WITH COUNTEREXAMPLES

You know the rules for adding and subtracting whole numbers so well that you hardly have to stop and think about them. In this lesson, you will use examples and counterexamples to explore statements about adding and subtracting with signed numbers.

Use Cubes to Model Calculations

How could you use cubes to model adding and subtracting positive and negative numbers?

Using cubes can help you get a better sense of how to calculate with positive and negative numbers. After the class discusses the handout Cube Calculations with Signed Numbers, complete the following:

1. On your paper, write each of the Signed Number Problems shown. Use cubes to figure out the solution. Then write the solution.
2. Make up a problem in which you need to add zero-pairs to your cubes to solve it. Start with 5 cubes. Illustrate your problem with cubes.
3. Make up a problem in which you need to add some negative cubes to solve it. Illustrate your problem with cubes.

Signed Number Problems

1. $6 + (-3)$	**2.** $-5 - (-4)$	**3.** $5 - (-3)$	**4.** $-4 + (-2)$
5. $-6 - 3$	**6.** $-3 - (-5)$	**7.** $5 + (-7)$	**8.** $-4 - 2$
9. $2 - (-1)$	**10.** $-4 + 7$		

Look for Counterexamples

For each statement below, decide if the statement will always be true. If the statement is not always true, show an example for which it is false (a *counterexample*). If the statement is always true, present an argument to convince others that no counterexamples can exist.

How can you argue that a mathematical statement is always true, or show that it is not always true?

1. I tried four different problems in which I added a negative number and a positive number, and each time the answer was negative. So a positive plus a negative is always negative. — Hyun

2. I noticed that a negative number minus a positive number will always be negative, because the subtraction makes the answer even more negative. — Tanya

3. I think that a negative number minus another negative number will be negative, because with all those minus signs, it must get really negative. — Hyun

4. A negative decimal number + a positive decimal number will equal 0 because they will cancel out. One example of this is −0.25 + 0.25. — Tanya

5. A positive fraction, like 3/4, minus a negative fraction, like −1/2 will always give you an answer that is more than 1. — Hyun

6. A negative decimal + a negative decimal will always give you a negative answer. — Tanya

7. You never need to add zero-pairs to your cubes when doing an addition problem — Hyun

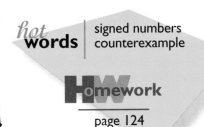

hot words | signed numbers counterexample

Homework
page 124

2 Counterexamples and Cube Combinations

PREDICTING RESULTS OF INTEGER ADDITION AND SUBTRACTION

In the last lesson, you explored statements about the results of adding and subtracting with signed numbers. In this lesson, you will analyze equations you create yourself. This will help you make predictions about whether the result of an equation will be positive or negative.

Use Cubes to Create Equations

How can you use cubes to create signed number problems when you know what the answer is?

The value represented by a given number of cubes depends on how many of the cubes are positive and how many are negative. In the following problems, build representations with cubes, and then record each equation in writing.

1 Make all possible combinations of 3 cubes. For each combination, bring in additional cubes so that the overall total is −4.

2 Make all possible combinations of 4 cubes. For each combination, remove cubes or bring in additional cubes so that the overall total is −1.

3 Make all possible combinations of 5 cubes. For each combination, remove cubes or bring in additional cubes so that the overall total is −1.

Signed Number Problem

Start with 6 negative cubes. Show a problem with the cubes for which the answer is −3.

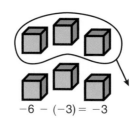

$$-6 - (-3) = -3$$

Sort the Solutions

Investigate the questions below about the results of adding and subtracting signed numbers. As you investigate, keep in mind that important question: Is it always true?

How can you predict whether an answer will be positive or negative?

1 What are the possible combinations of positive and negative numbers in an addition or subtraction equation that involves just two numbers? One example is positive + positive = positive. Be ready to share all the combinations you can think of in a class chart.

2 Look at the class chart of combinations.

   a. For which combinations on the chart can the sign of the result always be predicted?

   b. For which combinations on the chart does the sign of the result depend on the particular numbers in the equation?

Remember to look for counterexamples if you think you have found a rule.

Explain the Results

Choose one kind of combination from the class chart that will always have a negative answer. Choose another kind of equation from the class chart of combinations where the sign of the answer depends on the numbers being added or subtracted.

For both combinations, write one paragraph that explains the following:

- Will the result always be positive or negative? Why?

- If the result can be either positive or negative, what does it depend on?

hot **words** | signed numbers equation

Homework
page 125

3 More Cases to Consider

Now that you have learned how to add and subtract signed numbers, it is time to move on to multiplication and division. After finding the rules for multiplication, you will investigate rules for division. Then you will be ready to write your own statements and counterexamples to summarize what you have learned so far about signed numbers.

Use Cubes to Model Multiplication

How can you use cubes to model multiplication of signed numbers?

You have used cubes to model adding and subtracting with two numbers and made a chart of the results. Can you find ways to use cubes to model multiplying different combinations of signed numbers?

$$2 \times (-3)$$

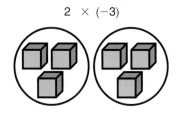

1. List all the different combinations of two signed numbers that could be multiplied.

2. For each combination you listed, write down some sample problems.

3. Try to find a way to use cubes to model examples of a multiplication problem for each combination of two numbers. Draw pictures of your models to record your work.

4. Write down any conclusions you can make about the results when multiplying each combination.

Investigate Division of Signed Numbers

Now that you have used cubes to model multiplying different combinations of signed numbers, the next step is to think about division. You may use cubes and what you know about the relationship between multiplication and division as you work on these division problems.

$$-6 \div (-3) = ?$$

1 List all the different combinations of two signed numbers that could be used in a division problem.

2 For each combination you listed, write some sample problems. Think about what you know about the relationship between multiplication and division before you write down the answers to your sample division problems.

3 Write down any conclusions you can make about the results when dividing each combination.

> **How can you use what you know about multiplication to think about division of signed numbers?**

Create Statements and Counterexamples

In your group, make up eight statements about operations with positive and negative numbers. Four of your statements should always be true. Four of your statements should not always be true and should have counterexamples. Try to include some statements that are true in some cases and others that are always false. Make sure you include all four operations in the eight statements: addition, subtraction, multiplication, and division.

On a separate sheet of paper, make an answer key for your statements that shows:

- the statement

- whether it is always true or not always true

- one counterexample, if the statement is not always true

hot **words** | signed numbers
counterexample

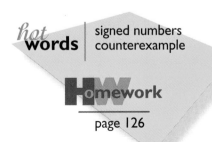

Homework

page 126

4 Rules to Operate By

In this lesson, you will think about rules for adding, subtracting, multiplying, and dividing with signed numbers. As you think about what operations might be equivalent and about counterexamples, you will be summarizing what you have learned about signed number operations in this phase.

Find Operations that Are Equivalent

Which operations with signed numbers are equivalent?

1 Using 3, −3, 5, and −5, write as many different addition and subtraction problems as you can that have the answer 2 or –2.

 a. Look at the problems you wrote and think about when adding and subtracting are equivalent, or when you get the same result.

 b. Write statements about when you think adding and subtracting are equivalent. Will the statements you have written always be true? Can you find any counterexamples?

2 Write down all the multiplication and division equations you can using all combinations of two numbers from 3, −3, 5, −5, 15, and −15. Use the equations you write to help you complete the following:

 a. Write a general rule for multiplication and division problems that have positive answers.

 b. Write a general rule for multiplication and division problems that have negative answers.

Find Counterexamples

Look at the statements in the box below.

▪ If you can find a counterexample, write the statement and its counterexample.

Is It Always True?
Positive + Positive = Positive
Positive + Negative = Negative
Negative + Negative = Negative
Positive − Positive = Positive
Positive − Negative = Positive
Negative − Positive = Negative
Negative − Negative = Negative

How can you apply what you know about integer addition and subtraction and counterexamples?

Write and Test Statements About Multiplication and Division

1 For each problem below, tell whether the answer will be positive or negative and how many negative numbers are multiplied.

a. $2 \times (-2)$

b. $2 \times (-2) \times (-2)$

c. $2 \times (-2) \times (-2) \times (-2)$

d. $2 \times (-2) \times (-2) \times (-2) \times (-2)$

2 Write a statement that tells whether your answer will be positive or negative when you multiply several numbers together.

3 Can you find a counterexample to the statement you wrote about multiplication? If so, rewrite your statement so that it is always true.

4 Using what you have learned about multiplication, write a statement that tells whether your answer will be positive or negative when the problem uses division.

hot **words** | signed numbers
counterexample

page 127

PHASE TWO

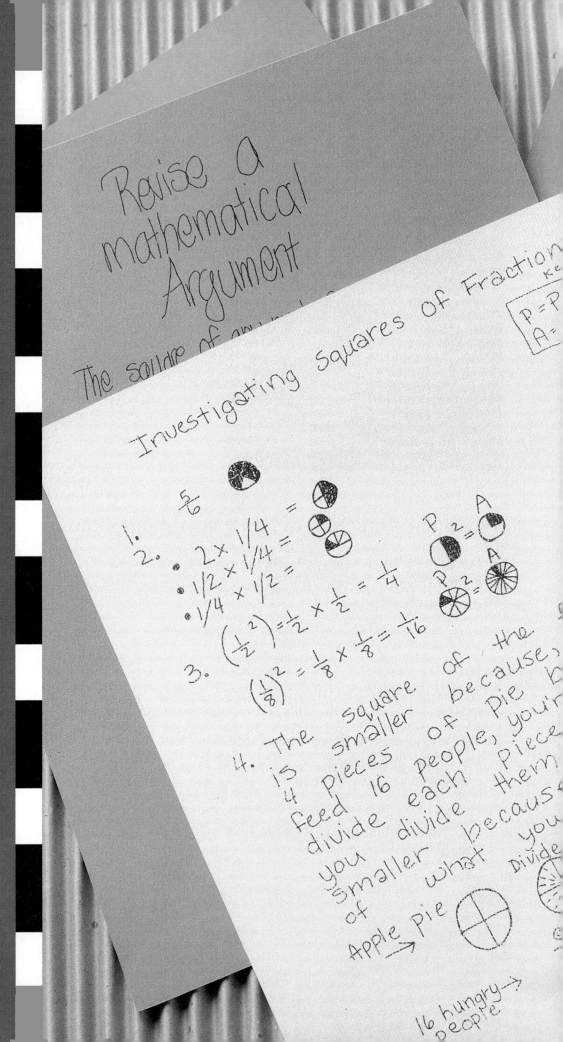

In this phase, you will be exploring patterns, squares, cubes, and roots. Testing rules for special cases will help you write your own mathematical arguments that apply to more general cases in mathematics.

Knowing how to represent squared and cubed numbers with drawings and cubes will help you understand more about them. How do you think you could show 4^3?

Roots, Rules, and Arguments

WHAT'S THE MATH?

Investigations in this section focus on:

NUMBER and RELATIONSHIPS

- Understanding how to find the square root, cube root, square, and cube of a number

- Investigating relationships among 0, 1, proper fractions, and negative numbers

- Understanding how to raise a number to an exponent

MATHEMATICAL REASONING

- Making mathematical arguments

- Evaluating mathematical arguments others have written

PATTERNS and FUNCTIONS

- Describing and finding perfect squares in tables and rules

MathScape Online
mathscape2.com/self_check_quiz

5 Perfect Pattern Predictions

EXPLORING
PATTERNS IN
PERFECT SQUARES

Any number multiplied by itself is called the square of that number. When the number that is multiplied by itself is an integer, the result is called a perfect square. In the last phase, you found rules for signed number operations. Can you find a rule to describe a pattern in perfect squares?

Investigate Perfect Squares

What is the pattern in the increase from one perfect square to another?

Look at the perfect squares shown in the box What's the Pattern? The number that is multiplied by itself to produce each square is called the square root. For example, the square root of 9 is 3. We can use a radical sign to write this as $\sqrt{9} = 3$.

1 The box What's the Pattern? shows the perfect squares that result when you square each whole number from 1 through 5. By how much is each perfect square increasing sequentially?

2 Make a table showing the increase when each whole number from 1 through 12 is squared. Label your table with three columns: Original Number, Perfect Square, and Increased By.

3 Extend your table to find between which two perfect squares the increase will be 21, 27, and 35.

What's the Pattern?

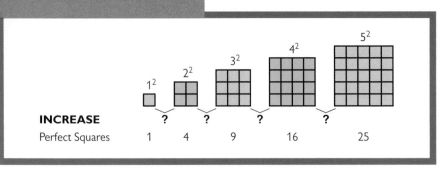

INCREASE

Perfect Squares   1   4   9   16   25

Find a Method to Predict the Increase

Come up with a method for figuring out the increase between any two perfect squares. You may find it helpful to use a calculator to investigate squares and square roots in coming up with a method.

1 Use words, diagrams, or equations to explain your method.

2 Think about how each step in your method relates to the way the perfect squares are shown with cubes. What do the numbers in your method represent in the cubes?

3 Use your method to figure out the increase between each of the following perfect squares and the perfect square that comes next: 529; 1,089; 2,401.

How can you predict the increase between any two perfect squares?

Write a Rule for the Pattern

Think about the method you came up with for predicting the increase for any perfect square.

- Write a rule that describes how to predict what the increase between perfect squares will be.

- Every positive square number has a positive and a negative square root. For example, 36 has square roots of 6 and -6 since $6^2 = 36$ and $(-6)^2 = 36$. -6 is the negative square root of 36. Does the rule you wrote work for the negative square root? If so, give examples. If not, give counterexamples and revise your rule.

hot **words** | perfect square square root

Homework

page 128

Counterexamples and Special Cases

In Lesson 5, you investigated patterns in perfect squares when positive and negative integers are squared. In this lesson, you will examine a mathematical argument about the result of squaring any number, looking for counterexamples and checking for special types of numbers. This will help you think of ways to revise the mathematical argument so that it is always true.

Find Counterexamples to Dan's Rule

How can you examine a mathematical argument to find counterexamples?

After you have read Dan's Mathematical Argument, work with your group to try to find counterexamples to Dan's rule. Work through some or all of these questions:

1 What do you think of Dan's rule?

2 Can you find any specific counterexamples for Dan's rule?

3 Can you find any types of numbers for which every number forms a counterexample?

4 What do you think is the smallest number for which Dan's rule is true? the largest number?

Dan's Mathematical Argument

What is my rule? My rule states: "The square of any number is always larger than the original number." For example, if I start with the number 8 and square it, I get 64. 64 is definitely larger than 8.

How did I figure out my rule? I started by choosing different numbers, like 3, 8, 12, 47 and 146. Each time I squared them, I got a larger number. I saw that as the original numbers got larger, the square numbers also got larger very quickly.

For what special cases is my rule true? I tried small and large numbers, like 3 and 146, and my rule was always true.

Original Number	Square Number
3	9
8	64
12	144
47	2,209
146	21,316

Revise a Mathematical Argument

How can you make a mathematical argument that is always true?

Think about Dan's Mathematical Argument and read it again if necessary. Write a new version of the argument that is correct and complete. Use the different points below as guidelines to make sure you have included all the information necessary to make your argument a strong one. Show all of your work.

1 What is your rule?

   a. Did you state your rule clearly?

   b. Could someone who had not already done the investigation understand your rule?

   c. Did you describe your rule generally so that it can apply to more than just a few numbers?

2 How did you figure out your rule?

   a. What methods did you use to figure out your rule?

   b. What counterexamples did you find where your rule did not work?

3 Does your rule apply to special cases?

   a. Does your rule work for 0? for 1? for fractions? for negative numbers?

   b. Are there other cases for which your rule works or does not work?

   c. If your rule does not work for some cases, explain why it doesn't.

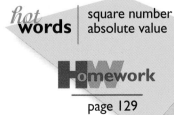

hot **words** | square number
absolute value

Homework

page 129

7 Root Relationships

You have already learned what a square root is and investigated mathematical arguments involving square numbers and square roots. In this lesson, you will explore cubes of numbers and cube roots. A perfect cube is a number that results when you use an integer as a factor three times.

Verify Dan's Mathematical Argument

What is missing from Dan's mathematical argument?

Follow the steps below to verify if Dan's argument is always true.

1 Check Dan's argument for counterexamples. List them and show why they do not fit Dan's rule.

2 Check Dan's argument for special cases. Is his rule true for 0, 1, fractions, and negative numbers? For each case, show the original number (or numbers) and describe how it compares to the cube of the number.

3 Do you have any special cases of your own that you want to check? Try them out and describe what you find.

Dan's 2nd Mathematical Argument

What is my rule? My rule states: "The cube of any number is always larger than the original number." For example, if I start with the number 3 and cube it, I get 27 and 27 is definitely larger than 3.

How did I figure out my rule? I started by choosing different numbers, like 2, 2.2, 3, 3.5, and 4. Each time I cubed them, I got a larger number. I saw that as the original numbers got larger, the cube numbers also got larger.

For what special cases is my rule true? I tried whole numbers and decimals, like 2 and 2.2, and my rule was always true.

Original Number	Square Number
2	8
2.2	10.648
3	27
3.5	42.875
4	64

Look at Square and Cube Roots

How do positive and negative numbers relate to cube and square roots?

As you answer the questions below, think about rules you might make.

1 Can you find each of these numbers? If you can, give an example. If you cannot, write "no."

  a. a positive number with a positive cube root
  b. a positive number with a positive square root
  c. a positive number with a negative cube root
  d. a positive number with a negative square root
  e. a negative number with a positive cube root
  f. a negative number with a positive square root
  g. a negative number with a negative cube root
  h. a negative number with a negative square root

2 What rules can you come up with for square and cube roots and positive and negative numbers? Why do your rules work?

Write and Revise a Mathematical Argument

Follow these steps to write and revise your own mathematical argument about cubes of numbers and original numbers.

1 Write a mathematical argument that includes a statement of the basic rule or mathematical idea, the methods used to figure out the rule, and any counterexamples you found where your rule did not work. Be sure to include a description of what happens in special cases.

2 Trade mathematical arguments with a partner. Look at the handout Guidelines for Writing Your Own Mathematical Argument.

  a. Did your partner meet all of the guidelines?

  b. Is there information in your partner's argument that you don't understand?

  c. Is there information you think your partner left out?

3 Revise your mathematical argument using your partner's feedback and any other information you have learned.

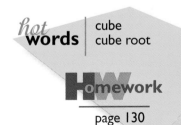

hot **words** | cube
cube root

Homework
page 130

8 A Powerful Argument

In this lesson, you will search for perfect squares in a chart which has numbers raised from the power of 2 to the power of 6. Exploring patterns in the location of these perfect squares helps you understand more about squares, roots, and exponents. This will prepare you for writing your own mathematical argument about one of the patterns.

Create a Powers Chart

What patterns of perfect squares can you find in The Powers Chart?

Make a chart like the one shown and follow the next three steps.

The Powers Chart

	$\square^1$	$\square^2$	$\square^3$	$\square^4$	$\square^5$	$\square^6$
1	1					
2	2					
3	3					
4	4					
5	5					
6	6					
7	7					
8	8					
9	9					

1 Use your calculator to fill in the chart by raising each number to the exponent shown at the top of the column.

2 When you have filled in all the numbers, use your calculator to check which numbers are perfect squares and circle them.

3 Do you see any patterns in the rows and columns where numbers are circled?

Write a Mathematical Argument

We have talked about a rule for the pattern found in the rows of the Powers Chart. This statement describes the pattern found in the columns of the chart: "If you raise any number to an even power, the result will be a perfect square."

1 Test the statement above and revise it by using the handout Guidelines for Writing Your Own Mathematical Argument.

2 Remember to include a statement of the basic rule, the methods used to figure out the rule and any counterexamples, and a description of what happens with special cases.

How would you write a mathematical argument about one of the patterns?

Test and Revise a Mathematical Argument

Work with a partner who will provide feedback and revise your mathematical argument by completing the following:

- Trade mathematical arguments with a partner.

- Read your partner's argument, checking to be sure your partner responded well to all of the questions in the Guidelines for Writing Your Own Mathematical Argument. Look for anything you don't understand and or anything your partner might have left out, and write down your comments.

- Return the mathematical argument and comments to your partner.

- Make revisions to your mathematical argument based on the written comments you receive from your partner.

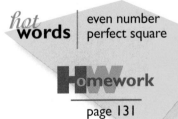

hot **words** | even number
perfect square

Homework

page 131

PHASE THREE

In this phase, you will write mathematical arguments about patterns involving divisibility, prime numbers, and factors. These topics have interested mathematicians since ancient times. A method for finding prime numbers was developed by a Greek astronomer around 240 B.C.

Today, powerful computers are used to search for prime numbers and prime numbers are used in computer codes for security.

Primes, Patterns, and Generalizations

WHAT'S THE MATH?

Investigations in this section focus on:

NUMBER THEORY

- Understanding divisibility, primes, factors, and multiples

MATHEMATICAL REASONING

- Making and understanding mathematical arguments

- Evaluating mathematical arguments others have written

PATTERNS and FUNCTIONS

- Using patterns to describe general rules

- Finding patterns in factors, primes, squares, or cubes

MathScape Online

mathscape2.com/self_check_quiz

Three-Stack Shape Sums

In the last phase, you used cubes to model square numbers. In this lesson, you model numbers by organizing cubes into stacks of three. Looking at numbers modeled in this way will help you investigate the statement: "The sum of any three consecutive whole numbers will always be divisible by 3."

Create Numbers Using the 3-Stacks Model

What patterns can you see when you make numbers with the 3-stacks model?

We say that one number is divisible by a second number if the first number can be divided evenly by the second number, leaving a remainder of 0. With the class, you have tried to see if you could find any three consecutive whole numbers whose sum is not evenly divisible by 3. Work with a partner and follow the investigation steps below to explore some patterns in consecutive numbers.

1️⃣ Use cubes to make each number from 1 to 15 following the 3-stacks model. On grid paper, record each number like the numbers shown in How to Use the 3-Stacks Model.

2️⃣ Look for patterns in the way the numbers look in the 3-stacks model. Be ready to describe any patterns you notice.

How to Use the 3-Stacks Model

L Number	b Number	Rectangle Number
4	5	6

Each time you have a stack of three cubes, start on a new column to the right.

Investigate the Sum of Numbers

In the last investigation, you found that numbers you made with cubes in the 3-stacks model could be described as L numbers, b numbers, or rectangle numbers. Follow the steps below to think about what it means to add any two kinds of numbers.

How can you predict whether the sum of two numbers will be an L number, a b number, or a rectangle number?

1 Make a table like the one shown. Put all the possible combinations of L numbers, b numbers, and rectangle numbers you can think of in the first two columns.

2 Then think about what your result would be if you added these kinds of numbers. Put that information in the third column.

3 What conclusions can you make about the sum of two numbers by looking at your table?

If you add this kind of number:	to this kind of number:	you get this kind of number:

Write and Revise a Mathematical Argument

Through class discussion, you have learned that a whole number is divisible by its factors. Write your own mathematical argument for this statement: "Whenever you add three consecutive whole numbers together, the sum will be divisible by 3."

- Remember to check your thinking. Are there any special cases you should consider? What happens with each of the special cases?

- Trade your argument with a partner and ask your partner to comment on it.

- Revise your mathematical argument based on your partner's comments.

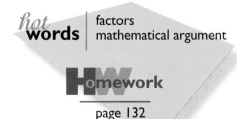

hot **words**
factors
mathematical argument

HW**omework**
page 132

10 A Stretching Problem

A prime number is a number that has exactly two factors, 1 and itself. In this lesson, you will be looking at patterns in prime numbers to solve a problem at a bubble gum factory. Then you will write your own mathematical argument about prime numbers.

Find the Unnecessary Machines

How can factors and multiples help you think about prime numbers?

First, read the information on this page about the Bubble Gum Factory to understand how it operates. Then read the handout The Bubble Gum Factory Script to find out what problem you can solve for the Bubble Gum Factory.

1 Look at the handout The Unnecessary Machines. Each one of the squares on The Unnecessary Machines is one of the machines in the Bubble Gum Factory. Some of the machines are unnecessary because combinations of other machines could be used instead.

2 Figure out which machines are actually unnecessary and cross them off. Be prepared to discuss with the class why you crossed off these machines.

The Bubble Gum Factory

At the Bubble Gum Factory, 1-inch lengths of gum are stretched to lengths from 1 inch to 100 inches by putting them through a stretching machine. There are 100 stretching machines. Machine 23, for example, will stretch a piece of gum to 23 times its original length.

Investigate the Necessary Machines

Use the questions below to find out how you could combine the *necessary* machines to get other lengths. Any of these machines may be used more than once to give the requested length.

What prime numbers can you use to make other numbers?

1 What machines could you use to get the lengths: 15? 28? 36? 65? 84?

2 For each of the lengths above, what other machines could have been used that were unnecessary?

3 Which lengths between 1 and 100 would come out if the bubble gum went through five machines and all 5 machines were necessary ones?

4 Which length between 1 and 100 requires the greatest number of necessary machines? How did you figure out your answer?

Write and Revise a Mathematical Argument

Write your own mathematical argument about this statement: "Any number can be written as the product of prime factors."

- Consider all the special cases we have used in this unit. Think about whether the special cases should be included in your argument. If a special case does not apply, it is sufficient to say so in your mathematical argument.

- Look for special cases other than 0, 1, proper fractions, and negative numbers. Your mathematical argument should describe the range of numbers for which the rule is true.

- Work with a partner to read and comment on each other's arguments.

- Revise your mathematical argument based on your partner's comments.

hot **words** | factors
prime number

Homework
page 133

11 Pattern Appearances

You will continue to think about squares, cubes, primes, and factors as you look for patterns in the Multiplication Chart. From these patterns you can make some general rules. Your goal will be to find out how many times any number would appear in the Multiplication Chart if the chart continued into infinity!

Find Out How Often Numbers 11–25 Appear

How many times will 11–25 appear on the Multiplication Chart?

After you work with the class to find out how many times the numbers 1–10 appear on the handout Multiplication Chart, expand your investigation by responding to the directions below. You will need the table you made with the class for this investigation.

1 How many times will 11 appear on the chart? 12? 13? Figure out how many times each of the numbers from 11 to 25 would appear on the chart and add them to your table. Remember to include how many times the number would appear if the chart went on into infinity, not just the number of times it appears on the Multiplication Chart you have.

2 How might you predict the number of times any number will appear as a product on the chart?

Multiplication Chart

×	1	2	3	4	5	6	7	8	9	10	11
1	1	2	3	4	5	6	7	8	9	10	1
2	2	4	6	8	10	12	14	16	18	20	2
3	3	6	9	12	15	18	21	24	27	30	3
4	4	8	12	16	20	24	28	32	36	40	4
5	5	10	15	20	25	30	35	40	45	50	5
6	6	12	18	24	30	36	42	48	54	60	6

Predict in Which Column a Number Belongs

How can you predict which column of your table a number belongs in?

You will need the table you made for this investigation. After exploring with the class which columns of your table the numbers 27, 36, 37, and 42 belong in, answer the questions below.

1 Choose five other numbers of your own between 26 and 100. What columns of your table do you think they belong in? Use the Multiplication Chart to check your predictions and then write the numbers in the appropriate columns on your table.

   a. How would you describe the numbers that appear in the 2-times column of your table?

   b. How would you describe the numbers that appear in the 3-times column of your table?

   c. Choose one other column of your table. How would you describe the numbers that appear in that column?

2 Identify numbers that belong in particular columns of your table.

   a. Can you think of a number that belongs in the 2-times column that is not already there? Write it in the column on your table.

   b. Can you think of a number that belongs in the 3-times column that is not already there? Write it in the column on your table.

   c. Can you think of a number that belongs in the column you chose in item **1c** that is not already there? Write it in the column on your table.

Generalize About the Column in Which Any Number Belongs

After the class discussion, write an answer to this question: How could you predict which column of your table *any* number would belong in?

- Be sure to include your own thinking about the question.

- Use examples to explain your answer.

hot **words** | pattern | factors

page 134

12 The Final Arguments

You will use what you have learned in this unit to write an argument about the numbers in the Multiplication Chart. You will also learn from others as you work together in small groups to share ideas about mathematical arguments.

Share Ideas About Mathematical Arguments

What will you write your mathematical argument about?

Choose something to write your mathematical argument about from the table you made for the handout Multiplication Chart. It could be one of the columns of your table, such as the 2× column, or it could be the numbers that appear most frequently. Your group may choose the same thing, or each person in the group could choose something different.

■ What is your rule for what you chose? How can you state your rule so that anyone who has not done the investigation yet will know what you are talking about?

■ Can you find any counterexamples to your rule? If so, how will you change your rule to take them into account?

■ Does your rule work for special cases? Why or why not?

Characteristics of a Good Mathematical Argument

A good mathematical argument should include the following:

- a rule that is general and clearly stated

- a description of how the rule was figured out, including a search for counterexamples (The rule should be revised if any counterexamples are found.)

- a description of special cases to which the rule applies

Write Your Own Mathematical Argument

Based on the discussion you had in your group, write your own mathematical argument using the following information:

1 Include all of the Characteristics of a Good Mathematical Argument on page 122.

2 Review the table you created and the Multiplication Chart from Lesson 11. Refer to the writing you did in Lesson 11 about how to predict which column of your table *any* number on the Multiplication Chart would belong in.

What do you include in a well-written mathematical argument?

Share and Revise Your Mathematical Argument

When you have completed your mathematical argument, trade with a partner.

- As you read your partner's argument, think of yourself as a teacher looking for a well-written mathematical argument. Write comments that you think will help your partner improve the argument.

- Use your partner's comments to revise your mathematical argument. Review the Characteristics of a Good Mathematical Argument again to make sure you have included all that is required.

hot **words** | mathematical argument
counterexample

Homework

page 135

Statements About Signs

Applying Skills

For items 1–8, a pink square represents a positive cube and a green square represents a negative cube.

1. What number do these cubes show?

2. What number do these cubes show?

3. What number do these cubes show?

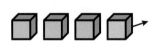

For items 4–8, use the pictures of cubes shown to solve a subtraction problem.

4. What number do the cubes show?

5. What has been added?

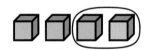

6. What has been removed?

7. What number will be shown by the remaining cubes?

8. What was the subtraction problem? What is the answer?

For items 9–13, solve each problem by drawing pictures of cubes.

9. $3 + (-1)$ **10.** $5 - (-3)$

11. $-2 + (-3)$ **12.** $-4 - (-1)$

13. $-6 - 4$

Extending Concepts

14. Make up a subtraction problem with a positive answer which can be solved using cubes, and for which you need to add some zero-pairs to your cubes. Draw the cubes to illustrate the problem.

For items 15–16, tell whether each statement is always true. If it is not always true, find a counterexample. Then rewrite the rule so that it is always true.

15. If you add a positive proper fraction and a negative proper fraction, you will always get a number smaller than 1.

16. If you subtract a positive fraction from a positive fraction, you will always get a number smaller than $\frac{3}{4}$.

Writing

17. Answer the letter to Dr. Math.

Dear Dr. Math,

My theory was: "The sum of a positive number and a negative number is always positive." I found 27 examples that worked. My friend Kate found one counterexample. So I figured 27 to 1, my theory must be good. But Kate said she only needed one counterexample to disprove my theory. Is this true?

Positively Exhausted

Counterexamples and Cube Combinations

Applying Skills

For items 1–8, solve each problem and write the entire equation.

1. $3 + (-2)$ **2.** $7 - (-5)$ **3.** $-8 + (-2)$

4. $-3 - 8$ **5.** $4 + (-2) + (-3)$

6. $-5 + 3 - 6$ **7.** $6 - 2 - (-1)$

8. $4 + (-2) - 3 + (-9)$

9. Each number in the tables below is found by subtracting the number in the top row from the number in the leftmost column. Copy each puzzle and fill in all the missing values.

−	−2	3	−4	8
1				
5		2		
−6				
0				

−	−3	4		
−2				
7		3	5	
				−3
			0	3

Extending Concepts

10. What combinations of four cubes can you come up with? For each combination, make up an addition or subtraction problem for which the answer is -4.

11. Find three different combinations of cubes, each representing the number -7. For each combination, make up a problem for which the answer is 4.

12. If you add a positive number and a negative number, how can you tell whether the answer will be positive, negative, or zero?

13. Suppose you start with the number -3 and add a positive number. What can you say about the positive number if the answer is positive? negative? zero?

14. If you add two positive numbers and one negative number, will the answer always be positive? If not, how can you tell whether the answer will be positive, negative, or zero?

Making Connections

The *emu* is a large flightless bird of Australia. The *paradoxical frog* of South America is so-called because the adult frog is smaller than the tadpole. Tell whether each statement in 15–19 is always true. If the statement is not always true, give a counterexample.

15. Some birds can't fly.

16. No bird can fly.

17. All birds can fly.

18. In every species, the adult is bigger than the young.

19. There is no species in which the adult is bigger than the young.

More Cases to Consider

Applying Skills

The multiplication 3 × 2 can be shown as 3 groups of 2.

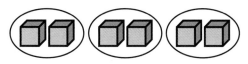

Draw cubes similar to the ones above to show each multiplication in items 1–3. Use shaded squares to represent negative numbers.

1. 2×4 **2.** $3 \times (-5)$ **3.** 4×3

Solve each problem in items 4–10.

4. $5 \times (-8)$ **5.** $6 \div (-2)$

6. $-9 \div (-3)$ **7.** $-8 \times (-7)$

8. $6 \times (-2) \div 3$ **9.** $-5 \times 4 \div (-8)$

10. $-4 \div 4 \times (-10) \div (-5)$

Extending Concepts

In items 11 and 12, look for different paths in the puzzles that equal some result. For each path you find, write an equation. You may move in any direction along the dotted lines, but you may only use each number once. Your path must start in the top row and end in the bottom row.

11. You may use only multiplication on this puzzle. A path that equals 24 is shown here.

Multiplication Puzzle

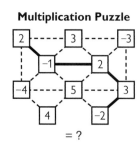

= ?

a. What path can you find that equals −240?

b. What is the longest path you can find? What does it equal?

12. In the puzzle below, you may use either multiplication or division. At each step, you choose which operation you want to use. You must use each operation at least once.

Multiplication and Division Puzzle

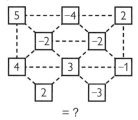

= ?

a. What path can you find that gives you an answer between −1 and −2?

b. What path can you find that gives you a positive answer less than 1?

Writing

13. Make up your own puzzle that uses multiplication, division, addition, and subtraction. Write two questions to go with your puzzle and make an answer key for each question.

Rules to Operate By

Applying Skills

For each subtraction problem in items 1–6, write an equivalent addition problem. For each addition problem, write an equivalent subtraction problem.

1. $8 - (-6)$ **2.** $9 - 2$ **3.** $7 + (-11)$

4. $10 - (-1)$ **5.** $3 + (-12)$ **6.** $4 + (-6)$

7. Using any pair of the numbers 5, −5, 8, and −8, write four different addition or subtraction problems that have the answer −13.

For items 8–9, use −6, −1, 2, −9, addition, and subtraction to solve each problem.

8. What problem can you find with the least possible answer?

9. What problem can you find for which the answer is zero?

10. Using the numbers $-6, \frac{1}{2}, -2$, and 4, and any three operations, what problem can you find for which the answer lies between −2 and 0?

Extending Concepts

To answer items 11–12 use the puzzle shown. You may use addition, subtraction, multiplication, or division. You must use each operation at least once, and your path must start in the top row and end in the bottom row. Remember, you may move in any direction along the dotted lines, but you may only use each number once.

11. What path gives an answer between 1 and 2?

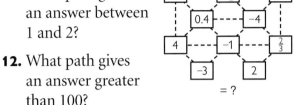

12. What path gives an answer greater than 100?

13. What statement could you write for which $2 + (-6) = -4$ is the counterexample?

Tell whether each statement in items 14–15 is always true or not always true. If it is not always true, find a counterexample.

14. If you subtract a negative fraction from a positive fraction, you will always get a number greater than $\frac{1}{4}$.

15. The product of two numbers will always be greater than the sum of the same two numbers.

Making Connections

For items 16 and 17, use the following information:

The Dead Sea is a salt lake lying on the Israel-Jordan border. At 1,292 feet below sea level, its surface is the lowest point on earth. At 29,028 feet, the top of Mount Everest is the highest point on earth.

16. Write a subtraction equation to find the elevation difference between the top of Mount Everest and the surface of the Dead Sea.

17. Write an equivalent addition equation.

Perfect Pattern Predictions

Applying Skills

For items 1–4, write each square using exponents and find their values.

1. 7 squared   **2.** 12 squared

3. 16 squared   **4.** 29 squared

For items 5–10, find each square root.

5. $\sqrt{49}$ **6.** $\sqrt{81}$ **7.** $\sqrt{196}$

8. $\sqrt{484}$ **9.** $\sqrt{1,024}$ **10.** $\sqrt{1,444}$

For items 11–14, find the increase between each pair of perfect squares.

11. 5th and 6th perfect squares

12. 11th and 12th perfect squares

13. 18th and 19th perfect squares

14. 30th and 31st perfect squares

For items 15–20, between which two perfect squares will the increase be:

15. 13?   **16.** 19?   **17.** 31?

18. 109?   **19.** 363?   **20.** 37?

Extending Concepts

For items 21–26, find each square root.

21. $-\sqrt{36}$ **22.** $-\sqrt{64}$ **23.** $-\sqrt{144}$

24. $-\sqrt{324}$ **25.** $-\sqrt{1,089}$ **26.** $-\sqrt{1,764}$

For items 27–29, use the pictures shown to answer the questions.

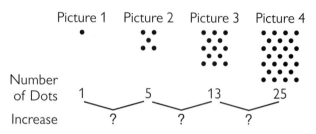

27. Find the increase in the number of dots between each of the figures shown. What pattern do you notice? What will be the increase in the number of dots between the 9th and 10th figures?

28. Describe a rule you could use to find the increase in the number of dots between any two figures. What will the increase be between the 100th and 101st figures?

29. The number of dots in the third figure is $3^2 + 2^2$ or 13. The number of dots in the 4th figure is $4^2 + 3^2$ or 25. What pattern do you notice? How many dots will be in the 40th figure?

Making Connections

30. The pyramid of Khufu in Egypt was built in 2680 B.C. as a burial tomb for the king. It has a square base measuring 756 feet on each side. The formula for the area of a square is s^2, where s is the length of a side. If each side of the pyramid was one foot longer, how much greater would the area of the base be?

Counterexamples and Special Cases

Applying Skills

For items 1–6, tell whether the square of each number is greater or less than the number itself. Do not calculate the square.

1. 5   **2.** 0.9   **3.** $\frac{1}{5}$ **4.** -18 **5.** $\frac{2}{3}$ **6.** -0.1

For items 7–9, show each multiplication by drawing a sketch, then give the result.

7. $\frac{1}{3} \times \frac{1}{2}$ **8.** $\frac{1}{2} \times \frac{1}{2}$ **9.** $2 \times \frac{1}{3}$

10. Copy and fill in a table like the one shown. Estimate the square root of each number in the table. Write down your estimate and then test it by squaring it. Repeat this process two more times. Then use your calculator to calculate how much your best estimate differed from the answer shown on the calculator.

	What is the square root of:	Estimate 1	Estimate 2	Estimate 3	By how much did your best estimate differ?
a.	43				
b.	86				
c.	306				
d.	829				

Extending Concepts

For items 11–14, consider the following rule: "The square root of a number is always less than the original number."

11. Give three counterexamples to the rule.

12. Test whether the rule works for each of these types of numbers: negative numbers, proper fractions, 0, 1, and positive numbers greater than 1. For types of numbers where the rule does not work, explain why.

13. Write a correct version of the rule.

14. Why do you think that 0 and 1 are often tested as special cases?

Writing

15. You learned in class that the square of a proper fraction is less than the original fraction. Use this theory to explain the meaning of counterexamples. You may want to give examples in your explanation.

Root Relationships

Applying Skills

Calculate items 1–6. To figure out the cube roots, you may want to use a calculator to guess and check.

1. 3^3 **2.** 11^3 **3.** $(-4)^3$

4. $(-9)^3$ **5.** the cube root of 125

6. the cube root of $-1,000$

For items 7–12, tell whether each number is a perfect cube.

7. 64   **8.** 16   **9.** -8

10. 25   **11.** 343   **12.** 1,728

For items 13–15, consider the following rule: "The cube of any number is greater than the square of the same number."

13. Give three counterexamples to the rule.

14. Test whether the rule works for each of these special cases: proper fractions, 0, 1, negative numbers, and positive numbers greater than 1. If there are some special cases for which the rule does not work, explain why not.

15. Write a new correct version of the rule.

Extending Concepts

16. Copy and fill in a table like the one shown. For each cube root, think about the two whole-number cube roots it might lie between. Write your answer in the second column. Estimate the cube root to the nearest hundredth. Write your answer in the third column.

	What is the cube root of:	Lies between cube roots:	Estimate
a.	50		
b.	200		
c.	520		

17. Is it possible to find a number with a negative cube root and a positive square root? If so, give an example. If not, explain why not.

Making Connections

For item 18, use the following:

Kepler's third law states that for all planets orbiting the sun, the cube of the average distance to the sun divided by the square of the period (the time to complete one revolution around the sun) is about the same.

18. Look at the table shown. Here is an example of the calculations for Earth:

$$d^3 = 93^3 = 804,357$$

$$T^2 = 365^2 = 133,225$$

$$\frac{d^3}{T^2} = \frac{804,357}{133,225} = 6.04$$

Is the law true for Earth, Venus, and Mercury?

Planet	Average distance to sun in millions of miles (d)	Period in days (T)
Earth	93	365
Venus	67	225
Mercury	36	88

A Powerful Argument

Applying Skills

Calculate items 1–10. To figure out the cube roots, you may want to use the calculator to guess and check.

1. 11^4 **2.** 7^3 **3.** 8^6

4. 4^5 **5.** 3^7 **6.** $\sqrt{529}$

7. the cube root of 729

8. $\sqrt{289}$ **9.** $\sqrt{1,225}$

10. the cube root of -1331

Without using a calculator, identify each number in items 11–19 that you know is a perfect square. Tell why you know.

11. 4^5 **12.** 5^6 **13.** 6^3

14. 8^7 **15.** 2^{10} **16.** 3^9

17. 9^3 **18.** 7^8 **19.** 11^4

20. Use your calculator to calculate each power in items 11–19.

21. Use the square root key on your calculator to find the square root for each number in items 11–19. Identify which of the numbers is a perfect square.

Extending Concepts

22. Copy the table shown. Which rows and columns in the table do you think are perfect cubes? How can you tell?

The Powers Chart

	$\square^1$	$\square^2$	$\square^3$	$\square^4$	$\square^5$	$\square^6$
6						
7						
8						

23. Complete the table by raising each number to the exponent at the top of the column.

24. Use guess-and-check and your calculator to check which numbers in the table are perfect cubes and circle them.

25. Someone made this mathematical argument: $x^a \times x^b = x^{a+b}$. What do you think about it?

Making Connections

Use this information for items 26 and 27.

Jainism is a religious system of India which arose in the 6th century B.C. and is practiced today by about 2 million people. According to Jaina cosmology, the population of the world is a number which can be divided by two 96 times. This number can be written in exponent form as 2^{96}.

26. Is 2^{96} a perfect square? How can you tell?

27. Is 2^{96} a perfect cube? How can you tell?

Three-Stack Shape Sums

Applying Skills

In the 3-stacks model, cubes are stacked in columns with a height of 3 cubes. The rightmost column may contain 1 cube (L numbers), 2 cubes (b numbers), or 3 cubes (rectangle numbers).

4   5   6

For items 1–6, draw a sketch showing how you could make each number using the 3-stacks model. Tell which kind of number each number is: a rectangle number, a b number, or an L number.

1. 7   **2.** 8   **3.** 12

4. 16   **5.** 17   **6.** 18

7. What kind of number is 22? 35? 41? 57? 71? 103? 261? 352?

For items 8–11, complete this question: What kind of number is the sum of:

8. an L number and a b number?

9. a b number and a rectangle number?

10. a rectangle number and an L number?

11. a b number and a b number?

Extending Concepts

For items 12–13, suppose the least of three consecutive integers is a b number.

12. What kind of number is the middle number? the greatest number?

13. What kind of number do you get if you add the two least numbers? all three numbers? Is the sum of the three numbers divisible by 3? How do you know?

In the 4-stacks model, cubes are stacked in columns with a height of 4 cubes. The rightmost column may contain 1, 2, 3, or 4 cubes as shown. Use the 4-stacks model to answer items 14–15.

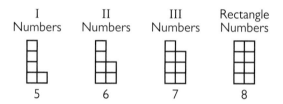

14. What kind of number do you get if you add a I number, a II number, a III number, and a rectangle number?

15. Is the sum of four consecutive integers always divisible by 4? Use your answer to item **14** to explain your answer.

Making Connections

16. The *mean* of a set of numbers is found by adding the numbers and dividing by the number of numbers. Do you think that the mean of three consecutive integers is always an integer? Use what you have learned in this lesson to explain your thinking.

A Stretching Problem

Applying Skills

For items 1–6, remember that at the Bubble Gum Factory, 1-inch lengths of gum are stretched to lengths from 1 to 100 inches by putting them through a stretching machine. There are 100 stretching machines. Which of the following machines are unnecessary?

1. 5   **2.** 14   **3.** 19

4. 31   **5.** 37   **6.** 51

For items 7–12, tell what necessary machines give these lengths:

7. 12 inches   **8.** 30 inches   **9.** 44 inches

10. 66 inches   **11.** 72 inches   **12.** 18 inches

For items 13–15, find the prime factors of each number.

13. 18   **14.** 48   **15.** 75

16. How can you tell whether a number is divisible by 3? by 9? by 5?

Extending Concepts

For items 17–19, use the same stretching machine described for items 1–12.

17. If you use only necessary machines, for which of the lengths 1 to 100 inches would exactly four runs through a stretching machine be needed? Give the lengths and tell which machines would be needed for each one.

18. Suppose a particular length can be obtained by using the necessary machines 2, 2, 3, 7. If you could also use unnecessary machines, what other combinations of machines could be used to obtain the same length? How did you solve this problem?

19. Describe a method to identify all the prime numbers between 1 and 200. Explain why your method works.

Writing

20. Answer the letter to Dr. Math.

> Dear Dr. Math,
> We had to find all the prime numbers between 1 and 100. I think I noticed a pattern: All the odd numbers that were *not* prime were divisible by either 3, 5, or 7. So now I can tell whether any number is prime: If it's even, I eliminate it right away. If it's odd, I have to figure out whether it's divisible by 3, 5, or 7. If it's not divisible by any of them, then I know I've got a prime number. Has anyone else noticed this pattern? How should I explain why this is true?
> In My Prime

Pattern Appearances

Applying Skills

Use the table you created in class or copy and complete the table below to show how many times each number from 1 to 25 would appear on a Multiplication Chart if the chart went on to infinity.

1 time	2 times	3 times	4 times	5 times	6 times	7 times	8 times	9 times	10 times
1	2	4							
	3								
	5								

For items 1–7, write down which column of the table each number belongs in.

1. 31   **2.** 26   **3.** 36   **4.** 41   **5.** 52   **6.** 58   **7.** 60

Extending Concepts

Use your table to answer items 8–11.

8. In which column would 132 appear? Why?

9. Find three numbers greater than 100 which belong in the 5-times column. How did you find them?

10. What can you say about the numbers that appear in the 2-times column?

11. Is it possible for a number that is not a perfect square to appear in the 3-times column? Explain why or why not.

For items 12–13, suppose you listed all the numbers from 1 to 100 in your table.

12. How many columns would you need? Which number or numbers would appear in the rightmost column?

13. Which column would have the most numbers in it? Why?

Making Connections

For items 14–15, use the following information:

The Ancient Egyptians had their own system for multiplying numbers. Examples of their system are as follows: To multiply a number by 2, double it. To multiply a number by 4, double it twice. To multiply a number by 8, double it three times. To multiply by 12, since 12 can be written as 8 + 4, multiply the number by 8 and by 4 and add the results.

14. What are the prime factors of 8? Why does it make sense that to multiply a number by 8, you would double it three times? Use this method to multiply 11 by 8 and describe the steps.

15. How do you think the Ancient Egyptians would have multiplied a number by 14?

The Final Arguments

Applying Skills

Use the table you created in class or copy and complete the table below to show how many times each number from 1 to 25 would appear on the Multiplication Chart if the chart went on to infinity. Use your table to answer items 1–5.

1 time	2 times	3 times	4 times	5 times	6 times	7 times	8 times	9 times	10 times
1	2	4							
	3								
	5								

1. The perfect squares 1, 4, 9, 16, and 25 all appear in a column headed by an odd number. Do you think that *all* perfect squares belong in a column headed by an odd number? If so, explain why you think so. If not, give a counterexample.

2. What is special about the perfect squares that appear in the 3-times column? Explain why this makes sense.

3. Are there any numbers in a column headed by an odd number that are not perfect squares? Why or why not?

4. Do you think that all perfect cubes, other than 1, belong in the 4-times column? If not, give a counterexample. What is special about the perfect cubes that appear in the 4-times column? Why does this make sense?

5. Summarize the patterns regarding perfect squares and perfect cubes in an accurate mathematical argument.

Extending Concepts

For items 6–8, test whether each statement about cube roots is true.

6. The cube root of any number is less than the original number if the original number is a positive fraction.

7. The cube root of any number is less than the original number if the original number is 0.

8. The cube root of any number is less than the original number if the original number is smaller than -1.

For items 9 and 10, test whether each statement about divisibility is true.

9. If you divide 10 by a positive fraction, the answer will be smaller than 10.

10. If you divide 10 by a negative number, the answer will be smaller than 10.

Writing

11. Write a paragraph explaining the steps that are involved in making an accurate and complete mathematical argument. Explain why each step is important. Which special cases might you test? Why?

To: Apprentice Architects
From: From the Ground Up.

Welcome to From the Ground Up.
As an apprentice architect, you
will be involved in all the stages
of designing model homes:
drawing floor plans, constructing
scale-size walls and roofs, and
estimating costs. To help you
design the homes, you will be
learning about geometry,
measurement, scale and
proportion, and cost calculations.

What math is involved in designing and building a model home?

FROM THE GROUND UP

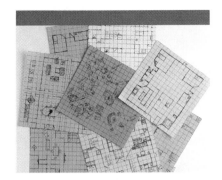

PHASE**ONE**
Floor Plans, Site Plans, and Walls

In this phase you will be using scale and metric measurement to design a floor plan, place the floor plan on a building site, and make outside walls with doors and windows. Next you will determine the area of the rectangular walls. Finally, you will calculate the cost of building the walls, based on area.

PHASE**TWO**
Roofs, Area, and Cost

You will experiment with ruler, scissors, compass, and tape to draw "nets" for four roof model types. After gaining expertise in roof design, you will design a roof for the house model you began in Phase One. You will end the phase by determining the area and cost of materials for your roof, floor, and ceiling.

PHASE**THREE**
Budgeting and Building

You will complete a cost estimate for your home design combining the estimated labor costs with the estimated costs for materials. By looking at costs for your classmates' model homes, you will see how cost and design are related. You will apply what you have learned in the unit to a final project: designing a home within a budget.

To: Apprentice Architects

Your first assignment is to plan and build a model home for the new community of Hill Valley. The Hill Valley developers want all the homes to be one story, but they also want a variety of floor plans so the houses don't all look alike. They don't want any rectangular floor plans.

We would like you to begin by designing a floor plan that meets Hill Valley's Building Regulations and Design Guidelines. We will soon be giving you more information about the walls, ceilings, roof, and costs. We're eager to hear your ideas for this project.

In this phase you will be designing a floor plan for a model home. By using scale, you will be able to represent the sizes of rooms and furniture on paper.

Making and understanding scale drawings and models is an important skill for many professions. Mapmakers, architects, and set designers all use scale. What other professions can you think of that use scale?

Floor Plans, Site Plans, and Walls

WHAT'S THE MATH?

Investigations in this section focus on:

SCALE and PROPORTION

- Figuring out the scale sizes of actual-size objects
- Figuring out the actual sizes of objects based on scale drawings
- Creating scale drawings

GEOMETRY and MEASUREMENT

- Estimating and measuring lengths
- Estimating and measuring areas of rectangles
- Comparing different methods for finding the area of rectangles

ESTIMATION and COMPUTATION

- Making cost estimates

MathScape Online
mathscape2.com/self_check_quiz

1 Designing a Floor Plan

CREATING AND
INTERPRETING
SCALE DRAWINGS

The first step in designing a model home is to create a floor plan. You will need to think about how to design a floor plan that meets both the Guidelines for Floor Plans and the real needs of people who might live in the home.

Relate Scale Size to Actual Size

What would your classroom look like in a scale where one centimeter represents one meter (1 cm:1 m)?

Comparing the Guidelines for Floor Plans to the room you are in and objects in it can help you get a better sense of the scale used.

1 Make a scale drawing of the perimeter of the classroom. On the drawing make an "X" to show your location in the classroom.

2 Draw a scale version of a teacher's desk and a student's desk in the scale 1 cm:1 m.

Guidelines for Floor Plans

- The home must be one story.

- Use a scale of one centimeter represents one meter to draw the floor plan and to build the walls and roof.

- The floor space must fit within a 16 meter by 16 meter square.

- The floor plan should *not* be rectangular.

Create a Floor Plan

Design a floor plan that meets the Guidelines for Floor Plans.

- Follow the Guidelines for Floor Plans on page 140.

- Include at least three pieces of furniture on your floor plan. Draw the furniture to scale to give people a better idea of what it might be like to live in this house.

How can you use scale to create a realistic floor plan?

Describe the Floor Plan

Write a description of your floor plan and what it shows.

- How did you decide how large and what shape to make the different rooms?

- How did you decide what size to make the scale drawings of the furniture?

- How can you check a floor plan to make sure that the spaces and sizes of things are reasonable for real people? (Is the bedroom large enough for the bed?)

Reflect on the math you used to design your floor plan.

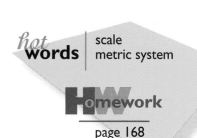

hot **words** | scale
metric system

HW**omework**

page 168

2 Site Plans and Scale Drawings

USING
PROPORTIONAL
RELATIONSHIPS IN
SCALE DRAWINGS

A site plan shows where the house will be placed on the building site. After placing your floor plan on the scale-size site to make a site plan, you will experiment with scale drawings and add some of them to your building site.

Make a Site Plan

How can you use what you know about scale to make a site plan?

To make a site plan, you will need your floor plan and an $8\frac{1}{2}$ in. by 11 in. sheet of paper that represents the scale size of your building site. Cut out your floor plan and glue it onto your building site. Make sure you follow Building Regulation #105 below.

- How large is your actual site in meters?

- How can you figure out the actual size of your site?

Guidelines for Floor Plans

Building Regulation #105:
The front of the home must be set back at least 5 meters from the street. The rest of the home may be no closer than 2 meters to the property line.

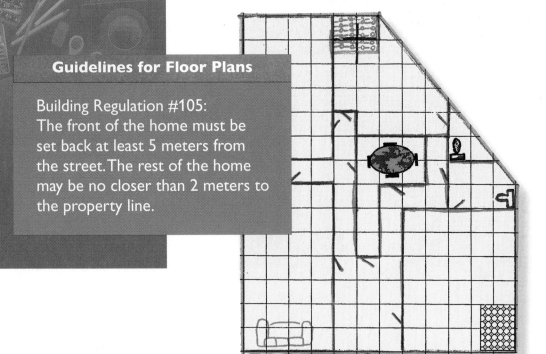

Experiment with Different Scales

Use the four different scales shown to make scale drawings of yourself or of another object, such as a tree, that is taller than you.

1 Show the front view of the person or object. (This is different from the top-down view you used in your floor plan.) You do *not* need to show details.

2 Cut out each drawing and label it on the back with the scale you used.

3 After you make the drawings, put them in order by size.

4 Glue the drawing you made in the scale of 1 cm:1 m to a card stock backing. Add tabs to the bottom so you can stand the drawing upright on your site plan.

How does changing the scale affect the size of a scale drawing?

Scales

1 cm:1 m

2 cm:1 m

$\frac{1}{2}$ in:1 ft

1 in:3 ft

Write What You've Learned About Scale

Write your responses to the following questions.

- Suppose you made many copies of the 1 cm:1 m scale version of yourself and then stacked the copies from head to toe. How many copies would you need to reach your actual height?

- What tips would you give to help someone figure out how large an object would be in different scales?

- Why do floor plans and site plans need to be drawn in the same scale?

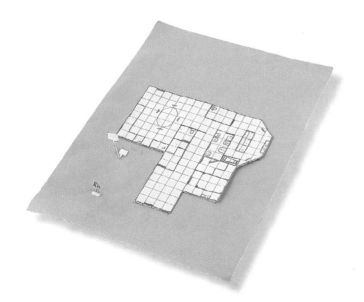

hot **words** | proportion
cross products

page 169

3 Building the Outside Walls

Choosing a realistic scale height is an important part of building the walls for your scale model home. To complete the walls, you will also need to think about how to make them match the perimeter of your floor plan.

Construct the Outside Walls

How can you build outside walls that are a realistic height and match the size and shape of your floor plan?

After deciding on your wall height, construct the outside walls around your floor plan. Be sure to follow the Guidelines for Walls.

- Experiment with methods for making the walls. Share methods that work well with your classmates.

- Keep in mind that there must be an exact fit between the outside walls and the perimeter of the floor plan.

Guidelines for Walls

- All outside walls of the house need to be the same height.

- All houses are to be one story.

- All houses must have a front door and back door for fire safety.

- Walls should be made with a scale of 1cm:1m.

Add Windows and Doors to the Walls

Now you are ready to add windows and doors to the walls. You will need to choose a reasonable size for the windows and doors on your scale model.

- Use your own method to add the windows and doors to your walls.

- Be sure to follow the guideline for doors in the Guidelines for Walls.

How can you use scale to decide how high up to place the windows?

How large should doors and windows be on the scale model?

Write About Walls, Windows, and Doors

Write about the thinking you used to make the walls, windows, and doors.

- What height did you choose for your outside walls? Explain why you chose that height.

- What suggestions would you give other students to help them construct walls for a model home?

- How did you decide what size to make the windows and how high to place them?

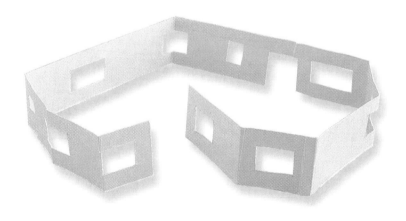

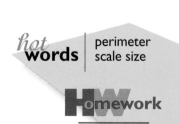

hot **words** | perimeter
scale size

Homework

page 170

4 Area and Cost of Walls

Estimating costs is an important part of any building project. After determining the area of walls for your model, you will be able to estimate the cost of building the actual-size walls. First, can you figure out the cost of a classroom wall?

Estimate the Cost of the Outside Walls

What would it cost to build the actual walls of your model house?

Use your house model to estimate the cost of building materials for the walls of the actual home.

1 Use whatever method you choose to determine the area and cost of the walls.

2 Record the cost on the Cost Estimate Form.

Cost Estimate Form

COST ESTIMATE FORM

	Model Area	Actual Area	Cost/m²	Total Cost	Estimate Checked by
Outside Walls	202.8 cm²	202.8 m²	$53.00	$10,748.40	
Roof					
Ground Floor					
Ceiling					
		Materials Subtotal			
		Labor at 60% of Materials			
		Total Cost Estimate			

...one By:

...Signature _____

...Checked By:

...nt's Signature _____

Cost of Building Materials for Walls

The cost of building materials for walls is $53 per square meter.

(For cost estimates, consider windows and doors to be part of wall area.)

Experiment with Wall Design and Costs

Refer to the Specifications for the Evergreen Project to create a scale drawing of a design for the front wall on centimeter grid paper. Use a scale of 1 cm:1 m. Show the size and position of the door and windows.

1 Use the rate of $53 per square meter to estimate the cost of the wall materials for the design you created. Consider doors and windows to be part of the wall area. Show your work.

2 Cut the cost of the front wall you designed by at least $350, but not more than $450. You can change the length and/or height of the wall, but not the $53 per square meter cost for materials. Describe your solution for cutting costs. Explain why your solution works.

3 Suppose you need to make a larger scale drawing of your wall design. The larger scale drawing should take up most of a 21.5 cm by 55 cm sheet of paper, and cannot be less than 50 cm wide.

   a. What scale should you use for the drawing? Explain your thinking.

   b. What are the dimensions of the wall in that scale?

How can you apply what you have learned about scale, area, and cost relationships?

Specifications for Evergreen Project

Measurements for Front Wall

Front wall: 4.5 m high by 13 m wide
Front door: 2.3 m tall by 1 m wide
Medium window: 1.5 m high by 1.2 m wide
Large window: 2 m high by 2.5 m wide

Guidelines for Walls

The wall needs to have a front door and 3 windows.
Use both medium and large windows.

hot **words** | area
rectangle

Homework

page 171

PHASE TWO

To: Apprentice Architects

Your next task is to make these four roof models: gable, hip, pyramid with rectangular base, and pyramid hip with pentagonal base of unequal sides.

You will design two-dimensional nets that will fold up to make each roof model. Nets aren't used for building actual roofs, but they are a great modeling technique. Each base should be about the size of your palm.

You will design a roof to match the floor plan of your model home. Then you will figure out the cost of building materials for an actual-size version of your model roof.

No house is complete without a roof. In this phase you will experiment with making nets for roof shapes and use a compass to design and construct a roof for your model.

To estimate the cost for roof materials, you will learn to find the area of the shapes that make up the roof. What shapes do you see in the roof nets pictured on these pages? What are the shapes of roofs in your neighborhood?

Roofs, Area, and Cost

WHAT'S THE MATH?

Investigations in this section focus on:

TWO- and THREE-DIMENSIONAL GEOMETRY

- Inventing and adapting techniques for constructing nets that will fold up into prisms and pyramids

- Constructing a triangle, given the lengths of its sides

- Using the compass to measure lengths and construct triangles

ESTIMATION and COMPUTATION

- Using a rate to make cost estimates based on area

AREA

- Building an understanding of area

- Finding the areas of triangles and trapezoids

MathScape Online
mathscape2.com/self_check_quiz

5 Beginning Roof Construction

Now you will experiment with making models of four different house roofs. To make each roof model, you will need to design a two-dimensional net that will fold up to make the roof shape.

Identify the Shapes of Roofs

What are the two-dimensional shapes that make up three-dimensional roof shapes?

Describe each roof in geometric terms. For each roof, identify the following:

- What is the three-dimensional shape of the roof?
- What are the two-dimensional shapes that make up the three-dimensional shape?

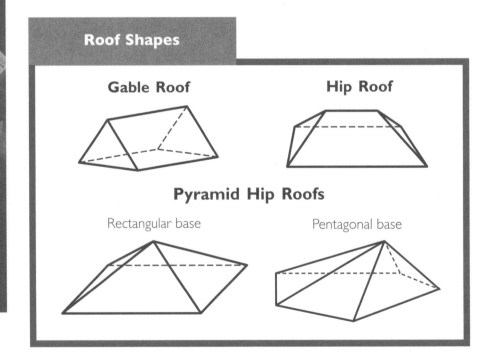

Roof Shapes

Gable Roof   **Hip Roof**

Pyramid Hip Roofs

Rectangular base   Pentagonal base

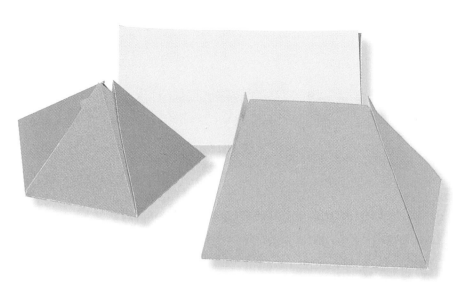

Experiment with Roof Models

Design a net that you can fold up to make a model of one of the four roofs. Cut out the net and fold it into shape.

- The base of the roof should be about the size of your palm.

- As you explore your own construction techniques, try to discover ways to get the parts of the net to match up so that the roof does not have any gaps.

How can you design two-dimensional nets to fold up into three-dimensional roofs?

Write About Construction Techniques

After you finish making a roof, use these questions to write about your experience:

- What was the most challenging part of designing your net?

- What tips would you give someone else?

- What are some differences between two-dimensional and three-dimensional shapes?

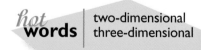

hot **words** | two-dimensional
three-dimensional

H W omework

page 172

6 Advanced Roof Construction

CONSTRUCTING
TRIANGLES WITH
COMPASS AND
RULER

The compass techniques on these pages will help you solve many of the problems you discovered as you explored roof construction. Use these techniques to make models of the four roof shapes shown on page 150.

Construct Triangles for Roof Nets

How can you use the compass to make constructions more precise?

You can use a compass and ruler to construct triangles for a net. The Building Tips below illustrate how to construct triangles for a hip roof net. After you construct the triangle for one side of the net, repeat the same steps to construct the triangle for the other side. Use the same method to make a gable roof net.

Building Tips

HOW TO USE A COMPASS TO CONSTRUCT TRIANGLES FOR A HIP ROOF NET

1. Measure edge *CD* by stretching the compass from vertex *C* to vertex *D*.

2. Keep compass at this setting and mark off an arc from vertex *C* and another from vertex *B*.

3. Where arcs cross is the vertex of the triangular face. Use a ruler to draw in the sides of the triangle.

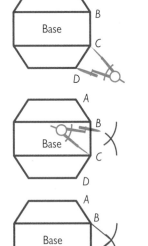

Construct a Pyramid Roof

You can use the following technique to make a pyramid roof over any polygon—rectangular or nonrectangular. Can you figure out why it works?

How can you use a compass to construct a pyramid roof?

Building Tips

HOW TO USE A COMPASS TO MAKE A PYRAMID ROOF

1. Draw the roof base and cut it out. Cut a straw to the roof height you want. Tape the straw vertically at about the center of the roof base.

2. Tape a piece of paper onto one edge of the base. Put the point of the compass at a vertex (*B*) of the roof base and stretch the compass to the top of the straw.

3. Keeping the point of the compass on *B*, draw an arc on the paper. Repeat Step 2 with the other vertex (*C*), and draw another arc.

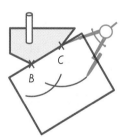

4. Use a ruler to draw lines connecting *B* and *C* to the point *D* where the arcs cross. This triangle is one roof panel.

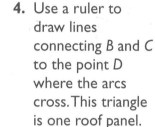

5. Cut out the roof panel. If you leave a tab on one edge, your roof will be easier to tape together.

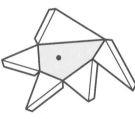

6. Continue to measure and lay out the rest of the roof panels. Remove the straw before you fold up and tape the roof.

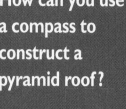

hot **words** | vertex
net

HW **omework**

page 173

7 Determining Roof Area

FINDING AREAS
OF TRIANGLES
AND TRAPEZOIDS

To find the cost of making your house model roof, you will need to find the areas of the shapes that make up the roof. After learning to find the area of a triangle by four different methods, you will consider how to find the area for a trapezoid.

Explore Triangle Area Methods

Why do different methods for finding the area of a triangle work?

On the handout Incomplete Net for a Hip Roof, experiment with each of the four different methods to find the area of the same triangular panel of the net. Write about why you think each method works.

Methods for Finding Triangle Area

Cover and Count

Cover the triangle with a transparent grid of centimeter squares. Count the number of whole squares plus the number of squares that can be made out of the leftover pieces.

Surround with a Rectangle

Draw a rectangle that surrounds the triangle and touches its vertices. Measure the length and width of the rectangle. Use the formula $A = lw$ to find its area. Divide the area by 2 to get the area of the triangle.

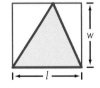

The Formula $A = \frac{1}{2}bh$

Draw in the altitude at a right angle to the base. Measure the base (b) and the height of the altitude (h). Use the formula $A = \frac{1}{2}bh$ to calculate the area.

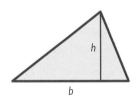

Draw the Midpoints Rectangle

Draw the midpoints rectangle. Measure its length and width. Use the formula $A = lw$ to find its area. Multiply the area by 2 to get the area of the triangle.

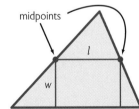

Find the Area of a Trapezoid

Look at the trapezoidal roof panels on the handout Incomplete Net for a Hip Roof. Which of the methods you used to find triangle area could you use to find the area of a trapezoid?

1 Find the area of the trapezoid in at least two different ways. Record the methods you used and the results. Explain why each method works.

2 Determine the total area of your hip roof model. Do not include the area of the roof base.

What methods can you use to find the area of a trapezoid?

Hint

FINDING THE FORMULA FOR THE AREA OF A TRAPEZOID

1. Label the trapezoid cutout as shown below.

2. Use a ruler to draw a diagonal line and cut the trapezoid into two triangles, as shown.

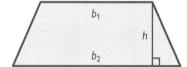
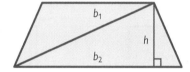

3. Can you arrange the two triangles to form one large triangle? Use what you know about the area of the large triangle to come up with a formula for the area of a trapezoid.

hot **words** | triangle
trapezoid

Homework
page 174

8 Calculating Roof Cost

ESTIMATING COST BASED ON AREA

Now you are ready to apply what you've learned about roof construction to make a roof for your model home. After estimating the area and cost of the roof, you will attach it to your model.

Design and Construct a Roof

How can you design a roof base, choose a roof style, and construct a roof for your model home?

Read the Guidelines for Roofs below.

1 Design the roof base and choose a roof style.

2 Use the techniques you learned on pages 150–153 to construct a net for the roof of your house model.

Sample
Floor Plan

Roof Base for
Pyramid Hip Roof

Rectangular
Roof Base

Guidelines For Roofs

All roofs must be sloped, not flat.

The roof base must overhang the floor plan by at least half a meter so that rain coming off the roof will drip down away from the walls.

If the perimeter of the floor plan is concave on any side, the roof base must be drawn convex to overhang that area. This will make it easier to construct the roof.

Find the Area and Cost of the Roof

Figure out the cost of the materials for your roof, based on the amount in the Cost of Roof Materials. Remember that the roof base should not be included in the roof area.

1 Draw a sketch of your roof and label each panel with a different letter (A, B, C, D, and so on). Find the cost of each panel. Record your work and label it with the panel letter.

2 Write about the method you used for finding area and why you chose to use that method.

3 On your Cost Estimate Form, fill in the model area, actual area, cost per square meter, and total roof cost.

How can you apply what you know about area to find the area and cost of the roof?

Cost of Roof Materials

The cost of roof materials is $48 per square meter.

Rate Your Model Home

Design

- Is the roof base a good fit for the shape of the floor plan?

- Is the shape of the roof a good match for the home?

- On a scale of 0 to 10, with 0 meaning "dull" and 10 meaning "innovative," how would you rate your home? Why?

Construction

- Do all sides of the roof match up precisely?

- Does the overhang follow the Design Guidelines?

- On a scale of 0 to 10, with 0 meaning "poorly crafted" and 10 meaning "masterpiece," how would you rate your home? Why?

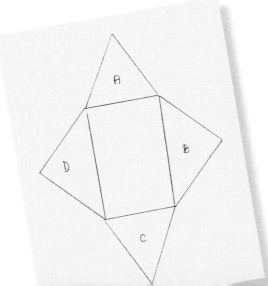

hot **words** | base
unit cost

Homework

page 175

PHASE THREE

Budgeting is an important part of building. To complete the cost estimate for your model home, you will need to find the area and cost of your floor and ceiling. Then you will need to add on the cost of labor.

By comparing and analyzing the costs of the different model homes completed in the last phase, you will learn some things about budgeting and building. This will prepare you for the final project: designing a new home, within a given budget, for a climate and building site of your choice.

Budgeting and Building

WHAT'S THE MATH?

Investigations in this section focus on:

SCALE and PROPORTION

- Making accurate drawings and scale models

GEOMETRY

- Planning and constructing three-dimensional models

AREA

- Applying methods for finding the areas of rectangles and triangles to finding the area of any polygon

- Planning and revising designs to fit within a given range of area measurements

ESTIMATION and COMPUTATION

- Planning and revising designs to fit within a given range of costs

MathScape Online
mathscape2.com/self_check_quiz

Costs of Floors and Ceilings

FINDING THE AREAS
OF POLYGONS TO
ESTIMATE COST

Your next step is to estimate the materials cost for the floor and ceiling of your house. The floor plans and ceilings fit into the geometric category of *polygons*. You will experiment with methods for finding the area for any polygon so that you can estimate the cost of your own floor and ceiling.

Investigate Polygon Area Methods

How can you find the area of any polygon?

Figuring out the area and cost of your floor plan and ceiling is a challenging task. Since they have irregular shapes, there's not one formula you can use for the areas. However, there are many different strategies that will work. See what you can come up with by experimenting with a sample floor plan.

1 What are the area and cost of the sample floor plan on the handout Sample Floor? Show your work. Keep notes about how you found the answers so you can report back to the class.

2 Describe how you would figure out the areas of the ceilings shown below.

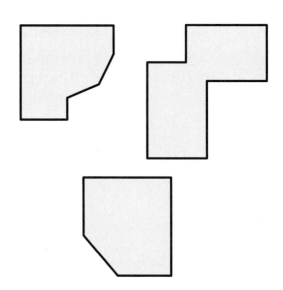

Materials Costs for Floors and Ceilings

- Floor materials cost $95 per square meter.

- Ceiling materials cost $29 per square meter.

Find the Floor and Ceiling Materials Costs

You will need a copy of your floor plan. To make a ceiling plan, make a copy of your roof base with overhangs.

1 What is the cost of materials for your floor and ceiling? Show your work.

2 Check to be sure the costs are reasonable. Then add the costs to your Cost Estimate Form.

How can you apply what you know about area to determine the cost of your floor and ceiling?

Compare Model Home Areas and Costs

Write about how the costs of the model homes compare.

- How does the area of your floor compare with the area of the ceiling, walls, and roof?

- How do the costs of your floor and ceiling compare with your classmates' costs?

- What are some ways to determine whether or not a cost estimate is reasonable?

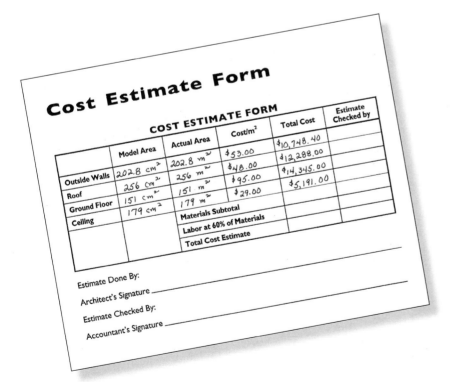

Cost Estimate Form

COST ESTIMATE FORM

	Model Area	Actual Area	Cost/m²	Total Cost	Estimate Checked by
Outside Walls	202.8 cm²	202.8 m²	$53.00	$10,748.40	
Roof	256 cm²	256 m²	$48.00	$12,288.00	
Ground Floor	151 cm²	151 m²	$95.00	$14,345.00	
Ceiling	179 cm²	179 m²	$29.00	$5,191.00	
		Materials Subtotal			
		Labor at 60% of Materials			
		Total Cost Estimate			

Estimate Done By:
Architect's Signature _____

Estimate Checked By:
Accountant's Signature _____

hot **words** | polygon
area

Homework

page 176

10 Adding the Cost of Labor

The final thing you need to consider for your model home is the cost of labor. Labor is people working to put the materials together into an actual house. How do you think the total cost of your model home will compare to the total cost of your classmates' models?

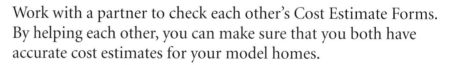

Find the Labor Cost and Total Cost Estimate

How can you find the cost of labor and estimate the total cost for your home?

Read the information on Cost of Labor below.

1. Figure out the cost of labor for your home.

2. Describe how you figured out the cost of labor. Fill in the amount on your Cost Estimate Form.

3. Figure out the total estimated cost for your home. Fill in the amount on your Cost Estimate Form.

Check Cost Calculations

Work with a partner to check each other's Cost Estimate Forms. By helping each other, you can make sure that you both have accurate cost estimates for your model homes.

Explain to your partner how you figured out the costs for your home. Then your partner will take on the role of Accountant and check all your calculations. If the Accountant finds any mistakes, you will need to make corrections on your Cost Estimate Form.

Cost of Labor

Labor costs are an additional 60 cents for every dollar of material costs. This means that labor costs are 60% of material costs.

Compare the Total Costs of the Homes

Copy the class data for all the homes.

1 What is the typical cost for the homes the class has made? (Find the mean and the median for the data.) Make a graph of the data on house costs.

2 What did you find out from the graph? Write a summary of the data.

3 How could you change your home so that it costs the median amount?

4 The bar graph below displays total cost data from the houses built by one class. How does your class's data compare?

How can you compare data on home costs?

Find the Square Unit Cost of Floor Space

Architects and realtors often use the cost per square unit of floor space to compare homes.

- How much does your home cost per square meter of floor space? For example, a $71,000 home with 155 square meters of floor space would cost about $458 per square meter of floor space.

- What information does the cost per square meter of floor space tell you about a house? How might that information be useful for comparing homes?

- How does your cost per square meter of floor space compare with the cost per square meter of your classmates' homes?

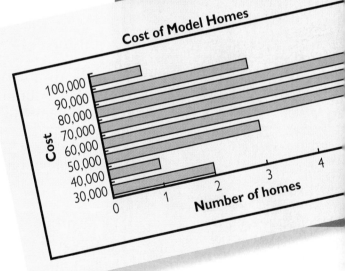

Cost of Model Homes

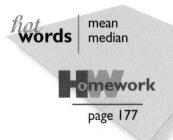

hot **words** | mean median

Homework
page 177

11 Designing Within a Budget

FINAL PROJECT

What have you learned about scale, geometry, area, measurement, and cost calculation? As a final project, you will apply what you have learned to design the best home you can. This time you will be designing for a certain climate and budget.

How will you design a home so that its cost is close to the target cost?

Make a Plan for the Final Project

Read the Final Project Design Guidelines. Choose one of the three building sites shown on this page and one of the four climate options on the handout Climate Options. Think about how the features of the land and the climate you chose will affect your home design.

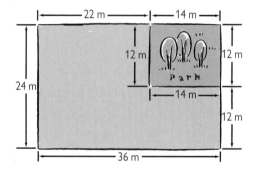

1 Make a scale drawing of your lot, using a 1 cm:1 m scale.

2 Write a brief description of your home.

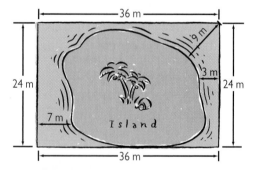

Final Project Design Guidelines

- The home should be suited to the chosen land and climate.

- The area of the ground floor of the home must be at least 150 square meters.

- The cost of building the home must be close to the target amount for the chosen climate.

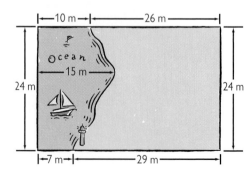

Work Independently on the Final Project

Use what you have learned about constructing floor plans, walls, and roofs, and about estimating costs to design the best home you can. Be sure to follow the Final Project Design Guidelines on page 164.

1 Design a floor plan. Draw a floor plan for the home using a scale of 1 cm:1 m. The home may have more than one story.

2 Make a rough estimate. Try to figure out if the cost of your home will be close to, but not go over, the target cost for your climate. If necessary, change your design.

3 Build a model. Use a scale of 1 cm:1 m. Make a carefully constructed model out of paper. Attach it to the lot.

4 Determine the cost. Figure out the cost of building your home. Show all the calculations and measurements you used in making your cost estimate.

Write About the Home

When you have finished building your home, write a sales brochure or a magazine article describing it. Include the information described on the handout All About Your Home.

How can you use what you've learned to design the best home you can?

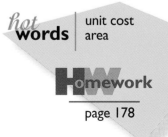

hot **words** | unit cost
area

Homework
page 178

12 Touring the Model Homes

It can be fun to see the different homes other people have designed for the same climate and budget as yours. How would you rate your final project? After you complete the House-at-a-Glance Cover Sheet and rate your own project, you will compare the homes.

Compare the Houses

How do the house designs, descriptions, and costs compare?

Make comparisons among the different houses your classmates have designed.

- What are the similarities and differences among homes for a particular climate?

- For a particular climate, how do homes with a low cost per square meter of living space compare with those with a high cost per square meter?

- What features do homes designed for similar building sites share?

- What different roof styles are used?

Reflect on the Project

- What do you like best about your home? What would you do differently if you designed another home?

- What mathematics did you use to design and build your house, figure its cost, and write your sales brochure?

- Suppose someone wants to build your house, but tells you it costs $10,000 too much. Describe how you might change your design to reduce the cost.

- Choose a different climate. What changes would you make to adapt your home to that climate?

- What does your project show about what you learned during this unit? What are some things you learned that your project doesn't show?

- Write a description of this unit for students who will use it next year. What math will they learn?

hot words | estimation
area

Homework
page 179

Designing a Floor Plan

Homework 1

Applying Skills

The scale drawings below are shown in a scale of 1 cm:1 m. Give the actual dimensions of each object. To do this you will need a metric ruler to measure the scale drawings.

1.

2.

3.

4.

Find the scaled-down dimensions of each room on a floor plan with a scale of 1 cm:1 m.

5. a bedroom that is 4 meters long and 3 meters wide

6. a kitchen that is 3.5 meters long and 2.8 meters wide

7. a living room that is 4.5 meters long and 5.75 meters wide

Tell what each total measurement would be in meters.

8. 23 centimeters

9. 80 centimeters

10. 5 centimeters

11. 1 meter and 15 centimeters

12. 2 meters and 85 centimeters

13. 1 meter and 8 centimeters

Extending Concepts

Make a scale drawing of each object, using an inch ruler and a scale of 1 in:1 ft. Show the top-down view.

14. a bathtub that is 4 ft 10 in. by 2 ft 4 in.

15. a piano with a base 4 ft 2 in. by 22 in.

16. a dresser with a base 31 in. by 18 in.

Find the dimensions of each room if it was drawn on a floor plan using a scale of 1 in:1 ft.

17. 12 ft 6 in. by 11 ft 6 in.

18. 13 ft 3 in. by 12 ft 9 in.

19. Estimate the length and width of your bedroom. Then find its dimensions in a scale of 1 in:1 ft.

Writing

20. Answer the Dr. Math letter.

> Dear Dr. Math,
> I drew the perimeter of my floor plan in a 1 cm:1 m scale. I want to use a different scale for drawing the furniture so that it will be easier to see. Is this a good idea? Why or why not?
>
> S. K. Ling

Site Plans and Scale Drawings

Applying Skills

Give the dimensions of each object using the scale indicated.

	Actual-Size Object	Scale
1.	1 ft by 3 ft	2 in:1 ft
2.	2 ft by 4 ft	2 in:1 ft
3.	3.5 ft by 5 ft	2 in:1 ft
4.	4 m by 3 m	2.5 cm:1 m
5.	1.5 m by 2.5 m	2.5 cm:1 m
6.	5.7 m by 13 m	2.5 cm:1 m

Give the height of a 5-meter-tall tree in each scale.

7. 3 cm:1 m

8. 1.5 cm:1 m

9. 1 cm:2 m

A basketball hoop is 10 ft tall. Give its height in each scale.

10. 1 in:3 ft   **11.** 2 in:1 ft   **12.** 1.2 in:1 ft

For items 13–14, refer to the scales below.

4 cm:1 m   1 cm:4 m   1 cm:2.5 cm

13. Of the scales listed above, which scale would result in the largest scale drawing of a chimney that is 9 m tall?

14. Which of the scales above would result in the smallest scale drawing of a chimney that is 9 m tall?

Extending Concepts

15. Measure or estimate the dimensions of two objects in your room. Write down their dimensions. What would the dimensions of each object be in a scale of 2 cm:1 m?

16. An apartment is 36 ft long and 25 ft wide. Give two scales you could use to make a sketch of the apartment floor plan that would fit on an $8\frac{1}{2}$ in. by 11 in. sheet of paper and take up at least half of the paper.

Making Connections

17. On average, Earth is 93 million miles from the sun. This distance is called one astronomical unit (A.U.). On a sheet of notebook paper, sketch a scale drawing that shows the distance of each planet from the sun, with the sun at one edge and Pluto at the other.

Planet	Average Distance from the Sun (A.U.)
Mercury	0.4
Venus	0.7
Earth	1.0
Mars	1.5
Jupiter	5.2
Saturn	9.5
Uranus	19.2
Neptune	30.1
Pluto	39.4

Building the Outside Walls

Applying Skills

Find the perimeter of each floor plan shown below. Assume that each square of the grid measures 1 m by 1 m.

1.

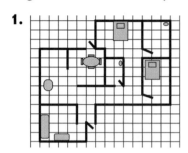

2.

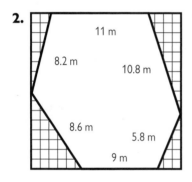

11 m

8.2 m

10.8 m

8.6 m

5.8 m

9 m

A sliding glass door is 84 inches tall and 72 inches wide. Give its dimensions on a wall plan using each of the following scales.

3. $\frac{1}{2}$ inch : 6 inch

4. $1\frac{1}{2}$ inch : 8 inch

5. $\frac{1}{2}$ inch : 1 foot

6. 3 inch : $1\frac{1}{2}$ feet

Extending Concepts

7. The rectangular lot for a model home has a perimeter of 100 cm. List the dimensions of as many rectangles as you can that have this perimeter. Use only whole-number dimensions.

8. For the model home in item 7, give three examples of rectangles with dimensions that are not whole numbers.

9. Suppose the average perimeter of a model house in a class is 75 cm and the wall height for all models is 2.5 cm. Estimate how many pieces of paper measuring 30 cm by 30 cm you would need for a class to build the walls of 32 model houses.

Making Connections

The type of glass used in windows is called *soda-lime glass.* It contains about 72% silica (crushed rock similar to sand), 15% soda, and 5% lime. Use this information to answer items 10 and 11.

10. How much silica, soda, and lime is there in 180 pounds of soda-lime glass?

11. What percentage of soda-lime glass is *not* silica, soda, or lime? Explain your reasoning.

Area and Cost of Walls

Applying Skills

Find the area of each rectangle. Assume that each square on the grid measures 1 cm by 1 cm.

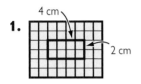

1. 4 cm ↘ 2 cm

2.

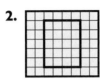

3.

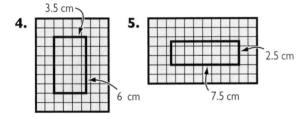

4. 3.5 cm ↘ 6 cm

5. 2.5 cm 7.5 cm

If materials for each square meter of outside wall cost $53, find the materials cost for each of the following.

6. an outside wall 2 m tall and 8 m long

7. an outside wall 2.5 m tall and 6.8 m long

8. the outside walls of a building with a perimeter of 60 m and walls 2 m tall

9. the outside walls of a building with a perimeter of 125 m and walls 2.4 m tall

Extending Concepts

10. Suppose the outside walls of a house are 2 meters high and the amount you can spend for outside wall materials cannot go over $7,500. For each material at the cost given in the table, what is the greatest perimeter the house could have?

Cost of Outside Wall Materials	
wood	$53 per square meter
stucco	$57 per square meter
brick	$65 per square meter
vinyl	$62 per square meter

For items 11–12, suppose you cannot spend more than $13,500 on outside wall materials for a house with a perimeter of 88 m.

11. For outside walls 2.5 m high, which of the outside wall materials listed in the table above could you afford?

12. Find the greatest wall height (to the nearest centimeter) that you could afford for outside walls of brick.

Making Connections

13. The Great Wall of China, dating from the third century B.C., is 2,400 kilometers long and from 6 to 15 meters tall. Estimate the area of one side of the Great Wall. Explain how you made your estimate.

Beginning Roof Construction

Applying Skills

For each figure below, name the three-dimensional shape. Name each two-dimensional shape that would be needed for a net of the three-dimensional shape. Sketch a net for the three-dimensional shape.

1.

2.

Tell whether each polygon would fold up to form a roof with no gaps. If it would, name the type of roof that would be formed.

3.

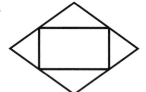

4.

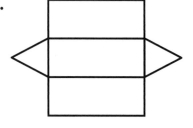

Extending Concepts

For each figure below sketch a net for the three-dimensional shape. Name each two-dimensional shape that appears in the net.

5.

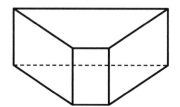

6.

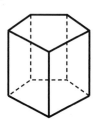

Making Connections

7. The Parthenon is an ancient temple in Greece. Imagine that you are building a scale model of the Parthenon. The width of your model is 32 cm and its length is 70 cm. The height of the gable at each end of the model is 6 cm.

   a. Sketch a roof net for your scale model. Give the dimensions of your net.

   b. Name the shapes you used.

8. Suppose the height of a column in the Parthenon is 10 m. Make a scale drawing of yourself standing beside the column. Tell the scale you used.

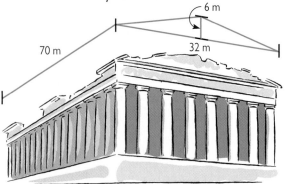

Advanced Roof Construction

Applying Skills

Use a compass and ruler to construct a triangle with the given side lengths.

1. 4 cm, 6 cm, 7 cm

2. 2 in., 2 in., 1 in.

3. 3 cm, 10 cm, 11 cm

4. 8.5 cm, 9 cm, 10 cm

5. 2 in., 2.5 in., 3 in.

6. 5 cm, 6 cm, 10 cm

Tell whether each set of side lengths can or cannot be used to make a triangle.

7. 2, 3, 4

8. 1, 2, 4

9. 11, 12, 14

10. 4.1, 5.5, 9.6

Extending Concepts

The chart gives the lengths of the sides of some triangles. Use what you know about triangles to give a possible length for the third side.

	Length of First Side	Length of Second Side	Length of Third Side
11.	5.3 cm	12 cm	
12.	30 m		40 m
13.		20 cm	30 cm
14.	15.04 m		15.04 m

For items 15–16, suppose that you have three 2-inch sticks, three 3-inch sticks, and three 5-inch sticks.

15. List all of the possible combinations of three sticks that can make a triangle. For instance, two 2-inch sticks and one 3-inch stick will make a triangle.

16. Suppose that you also have three 9-inch sticks. List all triangles that can be made with the 2-inch, 3-inch, 5-inch, and 9-inch sticks.

Making Connections

17. The Great Pyramid of Cheops was built in Egypt about 4,500 years ago.

a. Sketch a net for a model of this pyramid with a square base.

b. Which sides of your net would you measure out with a compass in order to make them the same length? Use an "X" to mark those sides on your net.

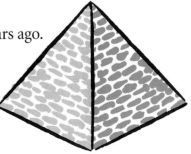

Determining Roof Area

Applying Skills

Find the area of each triangle or trapezoid.

1.

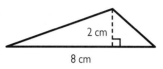

2 cm

8 cm

2.

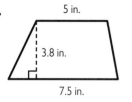

5 in.

3.8 in.

7.5 in.

3.

12 m

23 m

Find the area of each triangle.

4. $b = 5.8$ cm, $h = 16.2$ cm

5. $b = 88.6$ mm, $h = 23.4$ mm

Find the area of each trapezoid.

6. $b_1 = 7.0$ m, $b_2 = 9.7$ m, $h = 8.6$ m

7. $b_1 = 1.1$ m, $b_2 = 3.7$ m, $h = 0.6$ m

8. Find the area of the trapezoid.

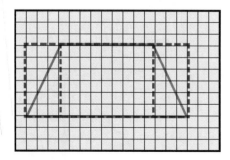

Each square represents one square centimeter.

Extending Concepts

9. In item 8, what is the area of the rectangle that surrounds the trapezoid? What is the area of the rectangle inside the trapezoid? How does the average area of the two rectangles compare to the area of the trapezoid? Will this be true for any trapezoid?

Making Connections

10. a. Find the areas of the triangle and rectangle shown.

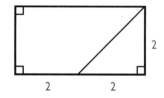

2

2   2

b. What percentage of the area of the rectangle is the area of the triangle?

c. Suppose that the areas of a second triangle and rectangle are in the same proportion and that the area of the rectangle is 40 cm^2. Set up and solve a proportion to find the area of the triangle.

Calculating Roof Cost

Applying Skills

Assume that each triangle or trapezoid below is part of a roof. Find its cost if roofing materials cost $51 per square meter.

1.

5.4 m

4 m

12.2 m

2.

3.11 m

7.25 m

3.

10 m

11.6 m

4.

19 m

11 m

24 m

If roofing materials cost $65 per square meter, find the cost of each roof. You may assume in each case that opposite faces are identical.

5.

3.4 m

1.9 m

3 m

6.4 m

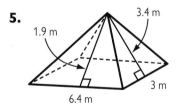

6.

2.2 m

2.4 m

2 m

4.6 m

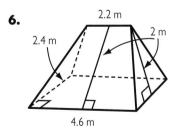

Extending Concepts

7. Two pyramid hip roofs have identical square bases. Each triangular face of the first roof has height 2.2 m. Each triangular face of the second roof has height 6.6 m. If roofing materials for both roofs cost $48 per square meter, how many times more expensive is the second roof than the first? Explain your reasoning.

8. When architects design a flat roof they must consider the weight of rain water the roof can support. If 1 m^2 on a roof is covered by water 1 cm deep, the water weighs 10 kg. Suppose a flat rectangular roof measuring 10 m by 20 m is covered by water 1 cm deep. How much does the water on the roof weigh?

Writing

9. Answer the Dr. Math letter.

> Dear Dr. Math:
>
> The architect showed me two possible designs for a house. The second house has less floor space and yet its roof is more expensive. My friend said he must have made a mistake because larger houses always have more expensive roofs. Is my friend right? If not, what might be the reason that the second house has a more expensive roof?
>
> A. N. Ewroof

Costs of Floors and Ceilings

Applying Skills

Find the area of each polygon.

1.

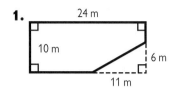

24 m
10 m
6 m
11 m

2.

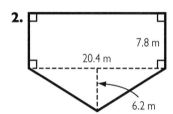

7.8 m
20.4 m
6.2 m

3.

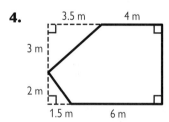

4.5 m
7 m
4 m
3 m
5 m

4.

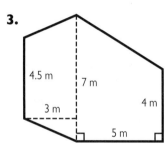

3.5 m   4 m
3 m
2 m
1.5 m   6 m

5. Find the cost of materials for the polygon in item **1** at $97 per square meter.

6. Find the cost of materials for the polygon in item **2** at $34 per square meter.

7. Find the cost of materials for the polygon in item **3** at $86 per square meter.

8. Find the cost of materials for the polygon in item **4** at $120 per square meter.

Extending Concepts

9. At a floor covering store, the prices for carpeting are given in square yards. The prices for vinyl floor tiles are given in square feet. Which would cost less: to cover the floor of your 10 ft by 12 ft bedroom in carpet costing $25 per square yard, or in tile costing $5 per square foot? How did you decide?

10. Suppose that your bathroom ceiling is covered with 9-inch-square tiles. You wish to remove the tiles and paint the ceiling instead.

   a. If you count 8 tiles on one side and 10 tiles going the other way, what is the area of the ceiling in square feet?

   b. Before painting the ceiling, you plan to install a new vent measuring 12 in. by 18 in. How much will it cost to paint the ceiling if paint costs $2 per square foot?

Making Connections

11. In a classic Japanese house, the floor is covered with tatami mats measuring 6 feet by 3 feet. The figure shows a floor plan for a Japanese house. Assuming that each rectangle represents a tatami mat, find the floor area.

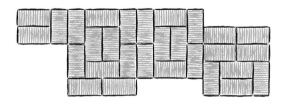

Adding the Cost of Labor

Applying Skills

If building materials cost $50,000, find the labor cost for each city.

1. Glenwood: 40% of materials cost

2. Duluth: 45% of materials cost

3. Bakersfield: 50% of materials cost

4. Sarasota: 65% of materials cost

Find the missing costs for each house if labor costs are 70% of the cost of materials.

	Cost of Materials	Cost of Labor	Total Cost to Build
5.	$39,560		
6.	$58,344		
7.	$71,875		

Check each calculation. First estimate whether the total cost is reasonable and tell how you decided. If the total cost is incorrect, give the correct answer.

	Cost of Materials	Cost of Labor	Total Cost to Build
8.	$43,000	50% of materials	$60,506
9.	$60,260	35% of materials	$81,351
10.	$50,390	60% of materials	$85,624

Extending Concepts

11. Suppose you are an estimator checking figures on a project. The cost of materials is $39,790. Labor costs are 53% of the materials cost. Estimate how much you would expect labor costs to be. Explain.

12. Better Builders Architects calculate the cost of labor by adding a fixed fee of $8,000 to 45% of the cost of materials.

    a. If materials cost $40,000, how much would labor cost? How much would labor cost as a percentage of the cost of materials?

    b. If the cost of materials is higher than $40,000, do you think that the cost of labor, as a percentage of the cost of materials, would be greater than, less than, or the same as your answer in part **a**?

Making Connections

13. This graph compares the average price per square foot to rent an office in four different cities. What would it cost in each city to rent an office with 800 square feet for 1 month?

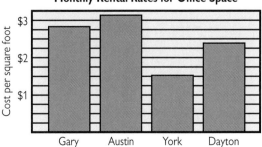

Monthly Rental Rates for Office Space

Homework 11

Designing Within a Budget

Applying Skills

Use the chart to find the total cost of materials for each house.

	Ground Floor	Roof	Outside Walls	Ceiling
1.	$9,920	$6,930	$8,750	$5,950
2.	$11,160	$8,227	$9,625	$7,350
3.	$15,500	$11,715	$10,780	$10,325

For the following houses, labor costs 55% of the materials cost. Tell whether each house is over or under the target total cost.

4. materials $49,204; target $75,000

5. materials $64,500; target $100,000

6. materials $101,320; target $155,000

7. materials $76,330; target $116,000

8. Suppose that you want to cut the cost of the house pictured in this floor plan. You do not want to change its total area or wall height. What can you do to keep the same floor area in your house at a lower total cost? Explain why your changes would reduce the total cost of the house.

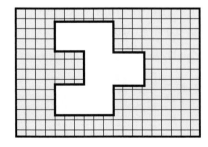

Extending Concepts

Refer to the rectangular floor plan for items 9–10.

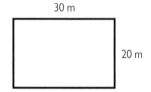

9. What happens to the area of the floor plan if you double the dimensions of all sides? Explain.

10. What happens to the area of the floor plan if you double only the length? If this is different than what happened in item **9**, explain why.

Making Connections

11. The lowest recorded temperature in Maybell, Colorado, is −61°F and in Tallahassee, Florida, it is −2°F.

   a. In which city would you expect material costs to be higher?

   b. If labor costs are about the same in both cities, in which city would you expect the cost of labor to be a lower percentage of the cost of materials?

Touring the Model Homes

Applying Skills

Find the cost per square meter of floor space to build each house. Round your answers to the nearest dollar.

		Total Area of Floor Space	Total Cost to Build
1.	Jose's house	84 m^2	$55,000
2.	Lee's house	175 m^2	$82,000
3.	Mei's house	620 m^2	$275,000
4.	Tom's house	105 m^2	$64,000

5. Considering total cost to build, order houses 1–4 from least expensive to most expensive.

6. Considering cost per square meter of floor space, order houses 1–4 from least expensive to most expensive.

7. This floor plan shows how the Lora family plans to add on to their house. The architect estimates that the addition will cost about $500 per square meter to build. What is the cost of the new addition?

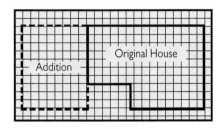

Each square represents one square meter.

Extending Concepts

8. Which surface would you choose for a driveway with an area of 300 ft^2?

Surface	Cost per ft^2	Lifetime
Brick	$7.50	20 years
Blacktop	$2.25	5 years
Concrete	$3.50	10 years

9. Read these ads. Which house gives you the most for your money? Tell why.

Move right in! Just $165,000 buys a 3-bedroom, 2-bath house on a large lot close to good schools and shopping. Large garden and garage. 1,760 square feet.

Clean and neat! Nice home in a tree-shaded neighborhood. 1,042 square feet; 2 bedrooms, 1 bath; new kitchen appliances and cabinets; carport. $95,960

Writing

10. Answer the Dr. Math letter.

Dear Dr. Math:

My dad says architects, builders, and real estate salespeople need to know lots of math. How do they use math in their jobs?

Lotta Mathews

THE
LANGUAGE
of
ALGEBRA

How can math help you describe and display relationships?

PHASE**ONE**
Writing Equations and Inequalities

In this phase, you will use variables to write equations and inequalities that describe your ideal school. You will also create a list of rules for deciding when two expressions are equivalent. Then you will have a chance to use these tools as you analyze data about possible school fund-raising events.

PHASE**TWO**
Creating Graphs of Equations

Here you will explore the coordinate plane as another way of describing things. You will move around a coordinate plane following directions given by your teacher. You will use graphing to help explore the relationship between Fahrenheit and Celsius. Finally you will make a graph showing potential profits from fund-raising.

PHASE**THREE**
Finding Solutions of Equations

What does it mean for a number to be a solution of an equation? First, you will explore the idea of a solution in a real-world situation. Then you will develop your own methods for solving equations. At the end of this phase, you will have a chance to use everything you have learned as you evaluate which company offers a better deal for field trip transportation.

PHASE ONE

In this phase, you will use variables to write equations and inequalities. These tools will help you describe your ideal school. They will also be useful when you analyze data about possible fund-raising events for a school.

Equation writing is an important ingredient in the language of algebra. Scientists often write equations to describe their work. What famous scientific equations can you think of?

Writing Equations and Inequalities

WHAT'S THE MATH?

Investigations in this section focus on:

ALGEBRA

- Using variables to write expressions, equations, and inequalities

- Graphing inequalities on a number line

- Making a table of values that satisfy an equation

- Deciding if two expressions are equivalent

DATA and STATISTICS

- Making a table of data

- Writing an equation to describe data

- Using an equation to make a projection

MathScape Online
mathscape2.com/self_check_quiz

 # Describing the Ideal School

How can mathematics be used to help describe the ideal school? You will start by setting up some variables and using them to write expressions. Then you can write equations to describe relationships about the school.

Write Expressions About a School

How can you use variables to write expressions?

You can put variables together with numbers and operations ($+$, $-$, $\times$, and $\div$) to write **expressions.** For example, the expression $b + g$ represents the number of boys plus the number of girls, or the total number of students.

1 Tell what each of the following expressions represents.

    a. $t + b + g$

    b. pl

    c. $(b + g) \div t$

2 Use variables to write an expression for each of the following.

    a. the total number of class periods per week

    b. the number of teachers who do not teach math

    c. the percentage of teachers who teach math

3 Write at least three new expressions that describe a school. You can use the variables listed above, the variables your class added, or completely new variables. If you use new variables, be sure to write down what they represent. Be prepared to share your expressions with the class.

Variables

A **variable** is a letter or symbol that represents a quantity. Here is some information about any school that can be represented by variables.

t = number of teachers at the school

m = number of math teachers at the school

b = number of boys at the school

g = number of girls at the school

p = number of class periods in one day

l = length of one class period, in minutes

Write Equations About the Ideal School

Imagine the school of your dreams. Work with a partner to describe the school using equations.

What equations can you write to help describe the ideal school?

1 Use the variables on page 184 or the variables your class added to write at least five equations that describe your ideal school. On a separate sheet of paper, make an answer key telling what the equations represent.

> 1. $m = 0.35t$
> 2.
> 3.
> 4.
> 5.

2 Trade your set of equations with another pair of students. (Hang onto the answer key!) For each of the equations you receive, write a translation of the equation into words.

3 For each equation, make a table showing some sample values that fit the equation.

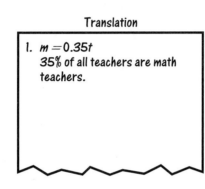

Translation

1. $m = 0.35t$
   35% of all teachers are math teachers.

Table

1. $m = 0.35t$
   35% of all teachers are math teachers.

t (teachers)	m (math teachers)
100	35
60	21

4 Trade back papers and see if your equations were interpreted correctly. Is there anything you would do differently? Be prepared to discuss this with the class.

Equations

An **equation** describes a relationship between two expressions. An equation tells you that two expressions are equal. Here are some examples.

$g = b + 43$ The number of girls equals the number of boys plus 43.

OR

There are 43 more girls than boys.

$m = \dfrac{1}{5}t$ The number of math teachers equals one-fifth of the total number of teachers.

OR

One-fifth of all teachers are math teachers.

hot words | expression
equation

Homework
page 212

2 Not All Things Are Equal

Equations are useful when you need to relate two expressions that are equal. When two expressions are not equal, the statements you can write are called inequalities. How can the language of inequalities help you say some more about your ideal school?

Express Inequalities in Different Ways

How many ways can you express the same inequality?

Inequalities can be translated into words and can be pictured on a number line. For each of the following, think of as many different ways of saying the same thing as you can. Use words, symbols, and pictures and keep a written record of your results.

1 $x < -2$

2 x is no greater than 4

3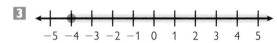

Inequalities

You can use inequalities to compare expressions that are not equal.

Symbol	Example	Number-Line Graph
 less than	$x < 3$ x is less than 3.	−5 −4 −3 −2 −1 0 1 2 3 4 5
≤ less than or equal to	$x \leq 3$ x is less than or equal to 3.	−5 −4 −3 −2 −1 0 1 2 3 4 5
> greater than	$x > 3$ x is greater than 3.	−5 −4 −3 −2 −1 0 1 2 3 4 5
≥ greater than or equal to	$x \geq 3$ x is greater than or equal to 3.	−5 −4 −3 −2 −1 0 1 2 3 4 5

Write Inequalities About the Ideal School

Imagine the school of your dreams. Work with a partner to describe the school using inequalities.

How can you use inequalities to describe the ideal school?

1 Use the variables on page 184 or the variables your class added to the list to write at least five inequalities that describe the ideal school. On a separate sheet of paper (the answer key), keep a record of what the inequalities represent.

1. $b + g \leq 1{,}500$
2.
3.
4.
5.

Answer Key
1. The number of students is less than or equal to 1,500.
2.
3.
4.
5.

2 Write at least five verbal descriptions of inequalities about the school. Then on the answer key, write these inequalities using symbols.

1. $b + g \leq 1{,}500$
2.
3.
4.
5. _____
6. There are at least 10 more girls than boys.
7.

Answer Key
1. The number of students is less than or equal to 1,500.
2.
3.
4.
5. _____
6. $b + 10 \leq g$
7.

3 Trade your set of inequalities and verbal descriptions with another pair of students. (Hang onto the answer key!) For each of the inequalities you receive, write a translation of the inequality into words. For each of the verbal descriptions you receive, write an inequality using symbols.

4 Trade back papers with the other pair of students and see if they interpreted your inequalities and descriptions correctly. Is there anything you would do differently? Be prepared to discuss this with the class.

hot **words** | inequality

Homework

page 213

3 Different Ways to Say the Same Thing

WORKING WITH
EQUIVALENT
EXPRESSIONS AND
EQUATIONS

Can two equations look different but say the same thing?
You will have a chance to write some equations that describe simple situations and see how your equations compare to those of your classmates. This will help you create a master list of rules for deciding when two expressions are equivalent.

Can you write more than one equation to describe a situation?

Write Equations from Situations

Write an equation to describe each situation. Then see if you can write another equation that also works.

1 Movable folding chairs in a school auditorium can be arranged in rows with an aisle down the middle. Here are two examples.

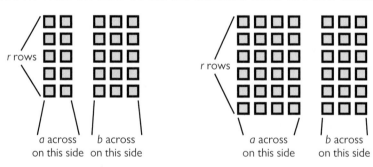

Write an equation for the total number of chairs, c, in the auditorium. Your equation should tell how c is related to r, a, and b.

2 A restaurant serves a buffet along a row of square tables. The row of square tables is surrounded by round tables for diners.

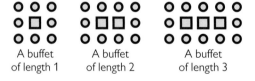

A buffet of length 1   A buffet of length 2   A buffet of length 3

Write an equation that tells how many round tables, T, there are for a buffet of length L.

Explore Rules for Equivalence

Work with classmates to write a list of rules that can be used to tell whether two expressions are equivalent. Try to state your rules as generally as possible. Begin by considering the following questions. (Remember that two equivalent expressions will give identical results no matter what values are chosen for the variables.)

1. Is $x + y$ equivalent to $y + x$?

2. Is $3(a + b)$ equivalent to $3a + 3b$?

3. Is $m - n$ equivalent to $n - m$?

4. Is $3x + 8x$ equivalent to $11x$?

5. Is $9y - 14y$ equivalent to $-5y$?

6. Is $a - a$ equivalent to 0?

7. Is $\frac{1}{2}x$ equivalent to $\frac{x}{2}$?

8. Is ab equivalent to ba?

9. Is $5 - x$ equivalent to $x - 5$?

10. Is $7(a - b)$ equivalent to $7a - 7b$?

Summarize the Rules

Write your own summary of the rules your class developed for deciding when expressions are equivalent. Include the following:

- a statement of each rule in words

- a statement of each rule using variables

- a specific example of each rule, or anything else that will help you remember how to use it

hot **words** | equivalent
equivalent expressions

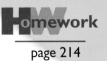
omework

page 214

What rules can you use to decide if two expressions are equivalent?

4 Raising Funds

How can you choose the best fund-raiser for your school?
An equation can help you decide.

Calculate School Revenue

How much profit can be earned from selling candy bars?

Woodmere School wishes to raise money to support their athletic teams. One option for fund-raising is to sell candy bars. You are on a committee formed to explore fund-raising and make a recommendation to the Student Body officers.

1 Examine the Candy Bar Fund-raiser. How much money can the school earn in profit by selling one box of candy bars? by selling 8 boxes?

2 Examine the candy bar data provided. Copy and complete the data table. How many boxes of candy bars must be sold for the school to earn $50 in profit? $100? $200?

3 Write an equation that relates the number of boxes of candy bars sold to the profit generated for the school.

Candy Bar Fund-raiser

The school buys candy bars by the box. Each box costs $7, including tax, and contains 12 candy bars. Shipping is free.

The suggested selling price for a candy bar is $1.

Candy Bar Data		
Boxes of Candy Bars	Total Revenue	Profit for School
1	$12	
2	$24	
3	$36	
4	$48	
5	$60	
10		
20		

Analyze and Compare Fund-raisers

Answer the following questions about the wrapping paper and magazine subscription fund-raisers.

1 How much will the school earn for each $6 roll of wrapping paper sold? for selling 20 rolls?

2 How many rolls would the school have to sell to earn $200?

3 Write an equation that relates the number of rolls sold to the profit to the school.

4 If you sell, on average, one magazine subscription weekly at the average price of $20, what would your total revenue be after 4 weeks?

5 How many subscriptions must be sold to make a profit?

6 What is the fewest number of subscriptions that can be sold to earn a profit of $50? $100? $200?

7 Write an equation that relates the number of subscriptions sold to the profit to the school.

What equations can help you describe and compare fund-raisers?

Wrapping Paper Fund-raiser

The school sells wrapping paper for $6 per roll.

The school will earn $\frac{2}{3}$ of the selling price for each roll sold.

Magazine Subscription Fund-raiser

Participants sell subscriptions for $20.

The company pays the school 50% of total sales in dollars.

The school must pay a flat fee of $15 for registration.

Write About the Results

Write one or two paragraphs to the Student Body officers summarizing the work of the committee. Include the following information in your report:

- How did you find the equations for each of the three products?

- How do the products compare as far as their potential for earning money for the school?

hot **words** | equation profit

page 215

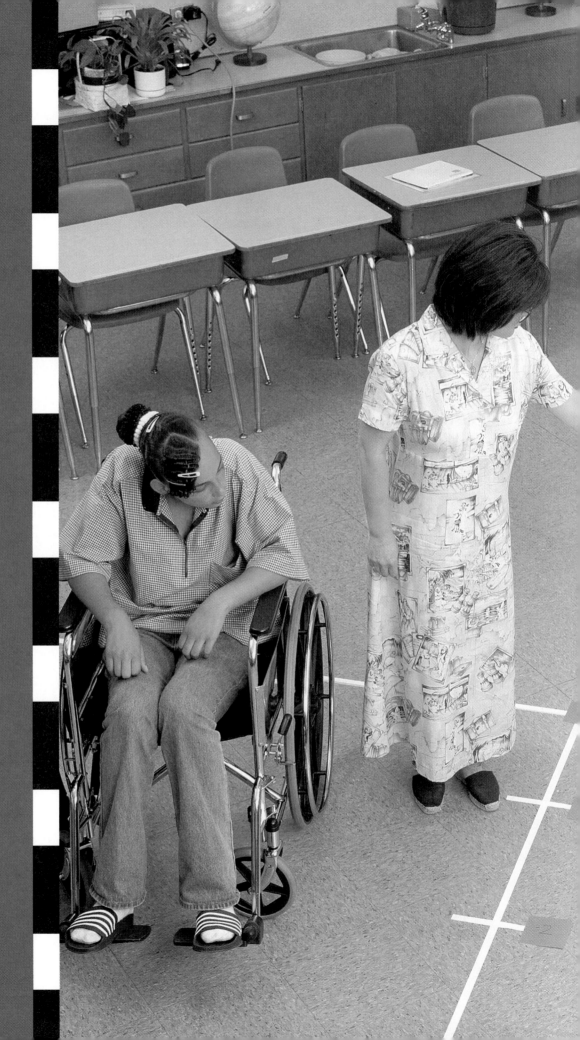

PHASE TWO

In this phase, you will explore the coordinate plane. You will see how the coordinate plane is connected to words, tables, and equations. By the end of this phase, you will be able to make a graph that shows potential profits from school fund-raising.

The coordinate plane is a grid that helps you describe the location of objects. What are some ways grids are used to describe locations in everyday life?

Creating Graphs of Equations

WHAT'S THE MATH?

Investigations in this section focus on:

ALGEBRA

- Plotting and naming points on the coordinate plane

- Making a graph of an equation

- Graphing horizontal and vertical lines

- Working back and forth among words, tables, equations, and graphs

DATA and STATISTICS

- Making a coordinate graph to display and analyze data

- Using a graph to make a projection

MathScape Online
mathscape2.com/self_check_quiz

5 Seeing Things Graphically

The coordinate plane is another important tool in algebra. Just like variables, tables, and equations, the coordinate plane gives you a way to describe and analyze situations. First you will explore the basics of the coordinate plane. Then you will see what happens when you plot points that fit an equation.

Explore Facts About the Coordinate Plane

What facts can you write about points on the coordinate plane?

What can you say about the coordinates of points that lie . . .

1 in the second quadrant?   **2** on the *x*-axis?

3 on the *y*-axis?   **4** to the left of the *y*-axis?

For each of the above questions, plot some points on the coordinate plane that fit the description. Write the coordinates of these points and ask yourself what the coordinates have in common. Write a short statement about each of your findings.

The Coordinate Plane

The **coordinate plane** is divided into four **quadrants** by the horizontal **x-axis** and the vertical **y-axis**. The axes intersect at the **origin**. You can locate any **point** on the plane if you know the **coordinates** for *x* and *y*. The *x*-coordinate is always stated first.

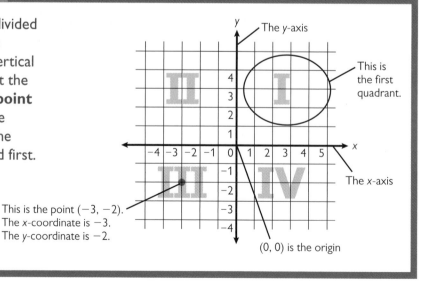

The y-axis

This is the first quadrant.

The x-axis

This is the point $(-3, -2)$.
The x-coordinate is -3.
The y-coordinate is -2.

(0, 0) is the origin

Plot Points from an Equation

Your teacher will give you an equation to work with for this investigation. Follow these steps to make a graph of your equation.

What happens when you plot ordered pairs that come from an equation?

1 Begin by making a table of values that satisfy the equation. You should have at least ten pairs of numbers, including some negative values.

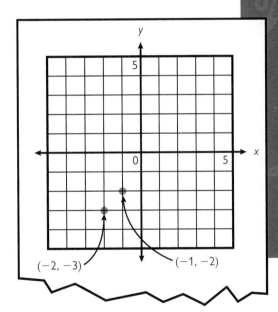

$y = x - 1$

x	y
−2	−3
−1	−2
0	−1
1	

2 Turn your table into a list of ordered pairs.

x	y	
−2	−3	→ (−2, −3)
−1	−2	→ (−1, −2)
0	−1	→
1	0	

3 Plot each of the points on the same coordinate grid.

(−2, −3)   (−1, −2)

4 What do you notice about your points? If you find more ordered pairs that satisfy your equation, do you think the new points will have the same property?

coordinates
ordered pair

omework

page 216

6 The Algebra Walk

Now you and your classmates will do a physical experiment to find out more about graphs. As you experiment, be sure to take notes on what you see. Then you will graph some equations by hand and compare results.

Do the Algebra Walk

What do you notice when your classmates move on a coordinate plane as directed by your teacher?

To do this experiment, your teacher will choose nine students to be "walkers." The other students are observers.

If you are a walker . . .

1 Stand at one of the integers on the *x*-axis that is marked on the floor. This is your **starting number.** Be sure to face in the direction of the positive *y*-axis.

2 Your teacher will give you directions on how to get a **walking number.** If your walking number is positive, move directly forward that many units. If your walking number is negative, move directly backward that many units. If your walking number is zero, stay put!

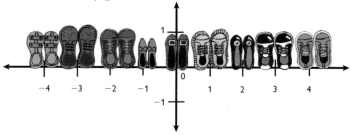

If you are an observer . . .

1 Watch carefully as your classmates move to their new positions on the coordinate plane.

2 Record their final positions with dots on your Algebra Walk Recording Sheet.

Graph Some Equations

For each of the following equations, make a table of values that fit the equation. Then make a graph of the equation.

1. $y = 2x$

2. $y = -2x$

3. $y = 2x + 1$

4. $y = 2x - 1$

5. $y = x + 1$

6. $y = -x + 1$

Compare the graphs you made to the results you recorded on your Algebra Walk Recording Sheet. Write a few sentences about what you notice.

How do the graphs of some equations compare to the results of the Algebra Walk?

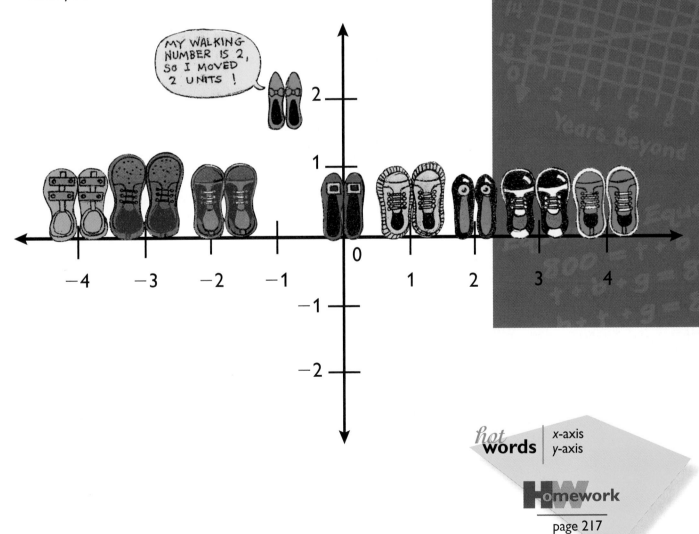

MY WALKING NUMBER IS 2, SO I MOVED 2 UNITS !

hot words | x-axis
y-axis

Homework
page 217

7 Putting It All Together

RELATING WORDS, TABLES, EQUATIONS, AND GRAPHS

How are words, tables, equations, and graphs connected?

First you will use these tools to explore horizontal and vertical lines. Then you will work with all four tools to explore the relationship between Fahrenheit and Celsius.

What do the equations of horizontal and vertical lines look like?

Explore Horizontal and Vertical Lines

What do equations, tables, and graphs look like when the lines they represent are horizontal or vertical? Work on the following items with a partner to help answer this question.

1 Make a table of values for the equation $y = 4$. Then create a graph for the equation.

2 Find the coordinates of at least five points on the line shown below. Use these coordinates to set up a table of x and y values. Look at your table. What do you notice? What equation do you think corresponds to this line?

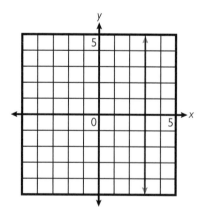

Show a Relationship in Four Different Ways

The equation $F = \frac{9}{5}C + 32$ describes the relationship between Fahrenheit and Celsius.

1 Write a description of the relationship between Fahrenheit and Celsius in words.

2 Make a table showing at least five pairs of values for Celsius (C) and Fahrenheit (F).

3 Make a graph that shows the relationship between Fahrenheit and Celsius. You will need to choose an appropriate scale for the axes on your graph.

A temperature of 0° Celsius is the same as a temperature of 32° Fahrenheit. How is this shown in each of your representations? Be ready to discuss this with the class.

How can words, tables, equations, and graphs help describe the relationship between Fahrenheit and Celsius?

Fahrenheit and Celsius

Temperatures are usually measured in degrees Fahrenheit or degrees Celsius.

Celsius   Fahrenheit

Water boils at 100° Celsius, which is the same as 212° Fahrenheit. — 100 — 212

Water freezes at 0° Celsius, which is the same as 32° Fahrenheit. — 0 — 32

hot **words** | horizontal
vertical

Homework

page 218

8 Presenting a Picture

MAKING GRAPHS TO DISPLAY AND ANALYZE DATA

In this lesson, you will use tables and equations to make graphs that help you analyze various fund-raisers. Then you will summarize what you know about words, tables, equations, and graphs—the four main tools of algebra.

Make a Graph of Fund-raisers

What can you learn from a graph?

In Lesson 4, you used tables and equations to describe three possible fund-raisers. The Student Body officers need to present your information to the school administrators, so they have asked you to make graphs showing the number of items sold and amount of profit for two of the products.

Choose two of the three products for more examination and analysis.

1 Review the equations you created in Lesson 4. What do the variables in your equations represent?

2 Let the *x*-axis show the number of units sold. The *y*-axis will show the profits to the school in dollars.

3 Choose a scale that will allow you to display the data you already have and to project how much profit might be earned if many sales are made. What do you need to consider about the scale if you want the two graphs to be easily compared?

4 Draw and label the axes. Remember to add titles to your graphs.

5 Plot some of the ordered pairs from the data tables in Lesson 4 for potential earnings and complete the graphs for the two products.

According to your graphs:

- How many of each item will need to be sold to raise $500?

- What is the profit if 30 units are sold for each item?

- Write some facts about the fund-raising that you learned when you made your graph.

Write About the Four Representations

You have seen four different ways to describe a situation: words, tables, equations, and graphs. These are called *representations*.

What can words, tables, equations, and graphs teach you about a situation?

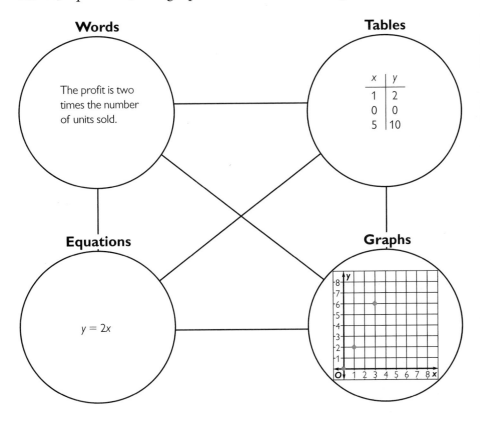

Words

The profit is two times the number of units sold.

Tables

x	y
1	2
0	0
5	10

Equations

$y = 2x$

Graphs

Write a summary of your work with the four representations. Include the following ideas:

- a specific example of when you used each representation,

- advantages and disadvantages of each representation, and

- a description of how you created one representation from another representation.

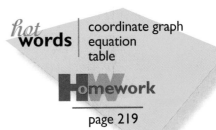

hot **words** coordinate graph
equation
table

Homework

page 219

PHASE THREE

Finding the solution of an equation is one of the most useful tools in algebra. In this phase, you will explore the idea of a solution. You will develop your own methods for solving equations and see how to use the solution of an equation.

At the end of this phase, you will examine the rate schedules of two bus companies. You will be able to write and solve equations to help you compare these companies and make a recommendation about which company the school should hire for a field trip.

Finding
Solutions of
Equations

WHAT'S THE MATH?

Investigations in this section focus on:

ALGEBRA

- Understanding what a solution is
- Checking the solution of an equation
- Finding solutions of equations using various methods
- Finding and graphing solutions of inequalities
- Applying solutions in real-world situations

DATA and STATISTICS

- Solving an equation to make a projection about data

MathScape Online
mathscape2.com/self_check_quiz

Situations and Solutions

How can the tools you have been exploring be used to solve problems? First you will solve a problem about rollerblade rentals. Then you will look more closely at what a solution is and use your own methods to find solutions.

How can you use tables, equations, and graphs to solve a problem?

Solve a Rollerblade Rental Problem

Refer to the advertisement for RolloRentals.

1 Make a table of values that shows the number of hours (*h*) and the corresponding cost (*c*) of a rollerblade rental.

2 Write an equation that describes the relationship between *h* and *c*.

3 Make a graph that shows the relationship between *h* and *c*.

Suppose you have $21 to spend. How long can you rent rollerblades from RolloRentals? Use your table, equation, or graph—or any other method you like— to help solve this problem.

An Advertisement

Solve Some More Rental Problems

For each amount of money, how many hours can you rent rollerblades from RolloRentals? (Note that RolloRentals will rent rollerblades for parts of an hour.)

1. $16

2. $28.50

3. $24.75

4. $106

How did you solve these problems? Write a short description of your method and be ready to discuss it with the class.

How can you find the number of rental hours for different amounts of money?

Another Advertisement

hot words | solution

Homework
page 220

10 Solving Simple Equations and Inequalities

DEVELOPING SOLUTION STRATEGIES

You have seen how to find a solution by making a table or graph. However, it is often useful to find a solution directly from an equation. You will explore some simple equations and develop your own methods for solving them. Then you will do the same for inequalities.

Find Solutions of Simple Equations

What methods can you use to solve simple equations?

Work with classmates to find a value of x that solves each equation. Use any methods that make sense to you, but be ready to share your thinking with the class. Be sure to check your solutions.

1. $6 + 2x = 20$

2. $42 = 10 + 4x$

3. $24 = 1.5x + 6$

4. $-4 + 3x = 11$

5. $3x = -18$

6. $-2x + 5 = 15$

Solutions of Equations

A **solution** of an equation is a value of the variable that makes the equation true.

For example, $x = 4$ is a solution to the equation $3x + 7 = 19$ because $3 \cdot 4 + 7 = 19$.

Write About Equation-Solution Methods

What method or methods can you use to solve equations like the ones shown above? Write a step-by-step description of the thinking you can use to solve these types of equations.

Find Solutions to Simple Inequalities

Work with classmates to find values of x that make each inequality true. Use any methods that make sense to you, but be ready to share your thinking with the class. Graph your solutions on a number line.

What methods can you use to solve simple inequalities?

1. $2x > 8$

2. $6 \leq 3x$

3. $x + 4 > 5$

4. $x - 1 \leq 7$

5. $x + 3 < 0$

6. $2x + 1 < 9$

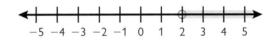

Solutions of Inequalities

A **solution** of an inequality is a set of values of the variable that make the inequality true.

For example, for the inequality $5x > 10$, any value of x greater than 2 makes this a true inequality. So, the solution is $x > 2$. This solution is shown on the number line.

$$\overset{\longleftarrow}{\underset{-5\ -4\ -3\ -2\ -1\ \ 0\ \ 1\ \ 2\ \ 3\ \ 4\ \ 5}{|\ \ |\ \ |\ \ |\ \ |\ \ |\ \ |\ \ \circ\!\!-\!\!|\ \ |\ \ |}}\overset{\longrightarrow}{}$$

Write About Inequality-Solution Methods

What method or methods can you use to solve inequalities like the ones shown above? Write a step-by-step description of the thinking you can use to solve these types of inequalities.

hot **words** | solution
inequality

Homework

page 221

11 The Balancing Act

Sometimes you can find the solution of a simple equation just by thinking about it logically. For more complex equations, it is helpful to have a method that you can rely on. You will begin by solving some balance problems. Then you will see how the idea of a balance can help you solve equations.

Perform a Balancing Act

How can you find a quantity that makes each scale balance?

Each picture shows a perfectly balanced scale. The cups contain an unknown number of marbles. In each picture, each cup contains the same number of marbles. Work with classmates to figure out how many marbles a cup contains in each picture.

1

2

3

Solve Some Equations

Solve each equation using whatever methods make the most sense to you. You might translate each equation into a balance picture and then find the number of marbles each cup must contain in order to keep the balance. Keep a step-by-step record of your work and check your solutions.

How can you use the idea of a balance to solve equations?

1. $2x + 6 = 3x + 2$

2. $8y + 2 = 2y + 14$

3. $25 = 7 + 3s$

4. $5m + 3 = 3m + 4$

Balance Pictures

Here's how to turn an equation (for example, $2x + 4 = 3x + 3$) into a balance picture.

1. Think of the variable as a cup containing an unknown number of marbles.

x is

2. Draw each expression as a set of cups and marbles.

$2x + 4$ is $3x + 3$ is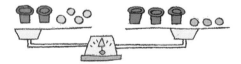

3. The equation says that the two expressions balance.

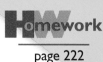

hot **words** | equation
solution

H-**W**omework

page 222

THE LANGUAGE OF ALGEBRA • LESSON 11   **209**

12

The Best Bus Deal

Applying your skills for analyzing data can help you make reasonable and appropriate decisions. In this lesson, you will use the four algebraic representations as tools for choosing affordable transportation.

Use Tables to Organize Data

How can tables and graphs help you make decisions?

Your school is planning to rent busses for a field trip to reward students who have performed community service. They will not know exactly how many students qualify for the field trip until the end of the semester, but the maximum number is 150 students. The field trip will last four hours. You are a member of the student committee that must investigate the costs of two possible bus companies.

Bus Company A	Bus Company B
The bus company generally used by the school district charges **$2.50 per student** and **$70 per hour**, regardless of the number of passengers or busses. Each bus can carry up to **30 students**.	The local charter bus company has offered to charge a rate of **$6.00 per student with no additional fees.** Each bus can carry up to **50 students.**

1 To evaluate the two companies, create a table of values for each company for a variety of passengers up to the maximum number of 150.

2 For each bus company, answer the following questions:

- What would be the cost if only 20 students go on the field trip? Write your answer in words and a number sentence.

- What would be the cost if 75 students were able to go?

- What would be the cost if the maximum number of students goes?

Represent Data With a Graph and an Equation

1 Create a graphical representation for each company.

- First you will need to decide what scale to use for the *x*- and *y*-axes.

- Then, plot the points representing your data.

- Show each company's graph on the same set of axes.

2 Write an equation for each company.

- What equations describe the cost for any number of students? What do the variables represent?

- Test your equations for 48 students. How much would each company cost?

Which bus company offers a better deal?

Write a Report

The students who have performed fifteen or more hours of community service will need transportation for the special reward field trip. You have been asked to present the information regarding bus costs to the school board.

You will need to write a short report recommending one of the two bus companies to the school board. To support your recommendation, include:

- multiple representations of the data you have found (tables, equations, graphs),

- an explanation of what information is highlighted by each representation,

- a description of how each representation contributed to your decision, and

- the advantages and disadvantages of using each company.

hot **words** | equation
solution

Homework
page 223

Describing the Ideal School

Applying Skills

These variables represent information about a particular math class.

s = number of students in the class.

g = number of girls in the class.

r = number of students in the class with red hair.

b = number of students in the class with black hair.

t = number of textbooks given to each student.

h = number of hours of classes each day.

Tell what each of the following expressions represents.

1. $s - g$ **2.** $b + r$

3. $5h$ **4.** st

5. $s - b$ **6.** $(g \div s) \times 100$

Write an expression for each of the following:

7. the number of students who do not have red hair

8. the number of textbooks handed out to girls

9. the number of minutes of classes each day

10. the number of textbooks handed out to boys

11. the percentage of students who have black hair

Write an equation that says . . .

12. there are 16 more students with black hair than students with red hair.

13. there are twice as many students as girls.

14. 40% of the students have black hair.

Translate each equation into words. Then make a table showing four pairs of sample values that fit the equation.

15. $s = b + 21$ **16.** $b = 4r$ **17.** $s = 2g + 5$

Extending Concepts

18. a. Write an equation that says that y is equal to 25% of x. Then make a table of values that fit the equation. For x, pick whole numbers ranging from 1 to 10. Write y as a decimal.

   b. Repeat part **a.** This time, use fractions instead of decimals.

19. Does it make sense for any of the variables listed at the top of the page to take values that are not whole numbers? If so, which ones?

Making Connections

20. The gravity on Jupiter is 2.64 times the gravity on Earth. This can be represented by the equation $J = 2.64E$. Make a table of values that fit the equation. If a person weighs 125 lbs on Earth, how much would the person weigh on Jupiter?

Not All Things Are Equal

Applying Skills

Tell which of the symbols $<$, $>$, or $=$ could go in the blank.

1. 4 ___ 9
2. 8 ___ 3

3. -8 ___ -6
4. -2 ___ 1

5. -0.1 ___ -0.2
6. 8 ___ 8

7. $\dfrac{2}{3}$ ___ $\dfrac{4}{6}$
8. -1 ___ 0

An algebraic inequality can be represented by words, symbols, or a number-line graph.

$x < 4$

x is less than 4.

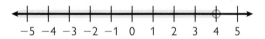

Use words to write each inequality. Express it in as many different ways as you can. Then make a number-line graph.

9. $x > 2$
10. $x < 5$

11. $x \le -1$
12. $x \ge -3$

Write each inequality using symbols and make a number-line graph.

13. x is greater than zero.
14. x is less than or equal to -3.

15. x is no less than 4.
16. x is less than 3.

17. x is no greater than -2.

Use words and symbols to write each inequality.

18.
```
←+——+——+——+——+——Φ——+——+——+——+——+——+——→
  −5  −4  −3  −2  −1   0   1   2   3   4   5
```

19.
```
←+——+——+——+——+——+——+——+——●——+——+——+——→
  −5  −4  −3  −2  −1   0   1   2   3   4   5
```

Extending Concepts

Suppose that g represents the number of girls in a particular class and b represents the number of boys. Write each inequality in items **20–22** using symbols. Then list all the possibilities for the number of girls if the number of boys is 20.

20. The number of girls is less than or equal to half the number of boys.

21. The number of girls is less than 2 more than the number of boys.

22. The number of boys is at least 3 times the number of girls.

23. Write four inequalities of your own using the variables g and b. Write each one using symbols and using words. Use each operation (addition, subtraction, multiplication, division) at least once.

Making Connections

24. As the moon goes around the earth, the distance between the earth and the moon varies. When the moon is closest to the earth, the distance is 227,000 miles. Use the letter d to write an inequality describing the distance of the moon from the earth.

Different Ways to Say the Same Thing

Applying Skills

Tell whether the two expressions in each pair are equivalent.

1. xy and yx 2. $3(a + b)$ and $3a + b$

3. $2x - y$ and $y - 2x$

4. $5(a - b)$ and $5a - 5b$

5. $x - 2y$ and $-2y + x$

6. $6(a + b) + a$ and $7a + 6b$

7. $2x + 3y$ and $5x$

8. $\dfrac{x}{y}$ and $\dfrac{y}{x}$ 9. $\dfrac{1}{3}x$ and $\dfrac{1}{3x}$

10. Which expressions are equivalent to the expression $2(x - y)$?

a. $2y - 2x$ b. $2x - 2y$

c. $2x - y$ d. $x + x - 2y$

11. Which expressions are equivalent to the expression $2a + 5(b - a)$?

a. $2a + 5(a - b)$ b. $5b - 3a$

c. $a + 5b$ d. $-3a + 5b$

12. Which expressions are equivalent to the expression $3x + 2y - x$?

a. $2x + 2y$ b. $2y + 2x$

c. $2(x + y)$ d. $4x$

Extending Concepts

13. The seats in a certain type of airplane are arranged as shown. The number of rows, r, and the width of the middle section, a, vary from plane to plane.

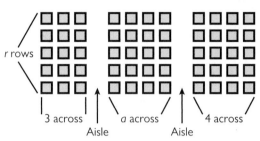

r rows | 3 across | Aisle | a across | Aisle | 4 across

a. Write an equation for the total number of seats, s, on an airplane. Your equation should tell how s is related to r and a.

b. Write at least two different equations equivalent to your equation in part **a**.

c. Check that your equations are equivalent by substituting values for a and r. Show your work. Use at least three different pairs for a and r.

Writing

14. Answer the letter to Dr. Math.

> Dear Dr. Math:
>
> I wanted to test the equivalence of the expressions $x(y - 1)$ and $xy - 1$. I decided to substitute values for x and y. I picked $x = 1$, $y = 2$. For both expressions, the result was 1. Then I picked $x = 1$, $y = 5$. For both expressions, the result was 4. So then I figured that the expressions must be equivalent. Am I right?
> P. Luggin

Raising Funds

Applying Skills

A phone company advertises that you can call anywhere in the United States for $0.04 per minute plus a monthly fee of $0.99 with their calling plan.

1. Make a table showing the monthly cost for five minutes, ten minutes, and twenty minutes.

2. Write an equation that relates the monthly cost c if you talk for a total of m minutes.

3. Use your equation to find the number of minutes if your monthly cost was $2.99.

The owner of a new restaurant is trying to decide how to place small rectangular tables to achieve maximum seating for large parties. All of the tables are the same shape and can be placed end to end to form a long table. One table can seat four people, two tables placed together can seat six people, and three tables placed together can seat eight people.

4. Make a chart showing how many people can be seated at four tables, five tables, and eight tables.

5. Describe how you can find the number of people who can be seated for any number of tables.

6. Write an equation that relates the number of tables x to the number of people y.

7. Use your equation to determine how many people may be seated at 12 tables.

8. If a party of 33 people came to the restaurant, how many tables would be needed to seat them?

Extending Concepts

The following data were provided by a fund-raising company. It shows the fund-raising efforts at five schools.

School	Units Sold	Revenue (dollars)	Profit (dollars)
A	135	877.50	351.00
B	97	630.50	252.00
C	314	2,041.00	816.40
D	150	975.00	390.00
E	100	650.00	260.00

9. Write an equation that relates the number of units sold to the revenue.

10. Write an equation that relates the number of units sold to the profit.

11. How much would a school earn in profit if it sold 213 units?

Writing

12. Write a summary of what you have learned in this phase about expressions, equations, variables, and inequalities. Be sure to explain the difference between an expression and an equation and the difference between an equation and an inequality. Also, describe how you can tell whether two expressions are equivalent.

Seeing Things Graphically

Applying Skills

1. For each point shown on the coordinate plane, give its coordinates and tell which quadrant it lies in.

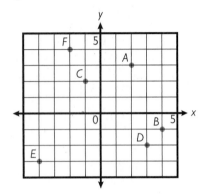

2. Plot each point on a coordinate plane.

a. $(3, 4)$ **b.** $(2, -1)$ **c.** $(-3, 1)$

d. $(-2, -4)$ **e.** $(3, 0)$ **f.** $(0, -1)$

Tell whether each statement is true or false. Explain your thinking.

3. The x-coordinate of the point $(-1, 3)$ is -1.

4. The point $(-1, 4)$ lies in the fourth quadrant.

5. The point $(-3, -8)$ lies in the third quadrant.

6. Any point whose y-coordinate is positive lies above the x-axis.

7. Any point whose x-coordinate is negative lies below the x-axis.

8. Any point that has a y-coordinate of 0 lies on the y-axis.

9. Any point whose x-coordinate is positive lies in the first quadrant.

Extending Concepts

10. a. Make a table of values that satisfy the equation $y = x + 3$.

b. Write the pairs of numbers from your table as ordered pairs.

c. Plot the points on a coordinate grid. Does it make sense to draw a line through the points?

d. Do you think that if you extended your line, the point $(100, 103)$ would lie on the line? Why or why not?

e. If a point lies on the line, what can you say about its coordinates?

Writing

11. Answer the letter to Dr. Math.

> Dear Dr. Math,
>
> My teacher asked for the coordinates of two points that lie on the x-axis. I figured that all points on the x-axis must have an x-coordinate of 0. So I wrote (0, 4) and (0, 98). My friend, Lou, said that I got it wrong. He said that points on the x-axis actually have a y-coordinate of 0. That sounds pretty silly to me. If Lou is right, why would they call it the x-axis? Who is right?
> Muriel

The Algebra Walk

Applying Skills

Suppose that in the Algebra Walk, the walkers calculate their walking number according to one of the instructions below.

A. Multiply starting number by 2 and add 1.

B. Multiply starting number by −2.

C. Multiply starting number by −2 and subtract 1.

D. Multiply starting number by 3.

Which set or sets of instructions are possible if . . .

1. the person with starting number 3 walks forward 7 units?

2. the person standing at the origin does not move?

3. after the Algebra Walk, nobody is standing at the origin?

4. the walkers with positive starting numbers walk backwards?

5. the students end up in a line that slopes downwards from left to right?

6. the student with starting number 2 walks 3 units farther than the student with starting number 1?

For each of the equations below, make a table of values that fit the equation. Then make a graph of the equation.

7. $y = 3x$

8. $y = -3x$

9. $y = 3x + 2$

10. $y = 3x - 2$

11. $y = 2x + 3$

12. $y = -2x + 3$

Extending Concepts

13. Write three equations whose graphs are lines that do not go through the origin. What do the equations have in common? Why does this make sense?

14. Write three equations whose graphs are lines that slant downwards from left to right. What do the equations have in common?

15. Write three equations whose graphs are lines that are steeper than the graph of $y = 5x$. What do the equations have in common?

Making Connections

16. When sound travels in air, its speed is about 12.4 miles per minute. Suppose that sound travels for x minutes. Let y represent the distance (in miles) that it travels. Then x and y are related by the equation $y = 12.4x$. Make a table of values that fit this equation and make a graph of the equation.

Putting It All Together

Applying Skills

Tell whether the graph of each equation is a horizontal line, a vertical line, or neither.

1. $y = 2x$ **2.** $x = 4$

3. $y = -5$ **4.** $y = -3x + 2$

5. $y = 7$ **6.** $x = 0$

Translate each equation into words. Then make a table of values and a graph.

7. $y = 3$ **8.** $x = -1$

9. $y = -4$ **10.** $y = 2x$

11. $y = 0$ **12.** $x = 5$

13. $y = -3x + 2$ **14.** $x = 0$

For each line in items **15** and **16** do the following: (a) Find the coordinates of five points on the line; (b) Make a table of x and y values from these coordinates; (c) Write the equation of the line.

15.

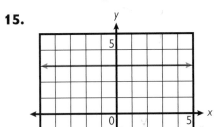

16.

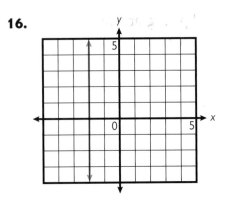

Extending Concepts

17. a. In addition to the equation on page 21, the relationship between Fahrenheit and Celsius can also be described by the equation $C = \frac{5}{9}(F - 32)$. Describe this relationship in words.

  b. Make a table showing at least five pairs of values for Fahrenheit (F) and Celsius (C).

  c. Make a graph of the equation. Show Fahrenheit temperature on the horizontal axis. Choose the scale on the horizontal axis so that Fahrenheit temperatures ranging from 0° to 100° are included.

  d. Pick a new point on the graph (one that is not included in your table in part **b**) and estimate its coordinates. Check the coordinates by using the equation. Explain how you did this.

Writing

18. Write a summary of what you have learned about coordinates, equations, tables, and graphs. Include an explanation of how you can graph the equation of a line. Also, describe how you can tell when the graph of an equation will be a horizontal or vertical line.

Presenting a Picture

Applying Skills

The figures below are formed using toothpicks. If each toothpick is one unit, then the perimeter of the first figure is 4 units.

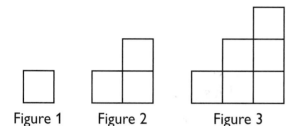

Figure 1   Figure 2   Figure 3

1. Make a table of the perimeter of Figure 1, Figure 2, Figure 3, Figure 4, Figure 5, and Figure 10.

2. Write an equation relating the figure number x to the perimeter y.

3. Draw axes and choose a scale for the x–axis and the y-axis. The x-axis should show the figure number. The y-axis should show the measure of the perimeter.

4. Plot the points from your table and make a graph of the perimeter of the figures.

5. Use your graph to predict the perimeter of Figure 100.

The figures below are formed by long strips of hexagon blocks, connected at one side. Each side of the hexagon blocks has a measure of one unit.

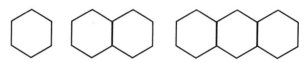

6. Make a table of the perimeter of figures made of 1 block, 2 blocks, 3 blocks, 4 blocks, 5 blocks, and 10 blocks.

7. Write an equation relating the number of hexagonal blocks x to the perimeter y. Explain how you determined the equation.

8. Draw axes and choose a scale for the x-axis and y-axis. The x-axis should show the number of hexagon blocks in the line. The y-axis should show the measure of the perimeter.

9. Plot the points from your table to make a graph of the perimeter of the blocks.

10. Use your graph to predict the perimeter of a figure made of 25 blocks.

Extending Concepts

11. How would your graph in item **9** be different if the measure of each side of the blocks was 2.5 units instead of 1 unit?

12. Does the change in item **11** describe a linear relationship? That is, will the perimeter continue to grow at a constant rate?

Situations and Solutions

Applying Skills

At his summer job, Rashad is paid a one-time bonus of $15 plus $8 per hour.

1. Make a table of values showing the number of hours Rashad works (h) and the corresponding total salary he receives (s).

2. Write an equation that describes the relationship between s and h.

3. Make a graph that shows the relationship between s and h.

4. How many hours would Rashad have to work to earn each amount?

   a. $63   **b.** $31   **c.** $67

   d. $35   **e.** $87   **f.** $47

Working at her father's office, Tonya is paid a one-time bonus of $20 plus $9.50 per hour.

5. Repeat items **1–3** for Tonya's job.

6. How many hours would Tonya have to work to earn each amount?

   a. $58   **b.** $67.50   **c.** $43.75

   d. $48.50   **e.** $77   **f.** $81.75

Extending Concepts

7. a. Describe a general rule you can use to figure out the number of hours Rashad must work to make any given amount of money.

   b. Describe how you can use your graph in item **3** to figure out the number of hours Rashad must work to make any given amount of money.

   c. Suppose you want to know how many hours Rashad must work to make $44.60. Can you find an exact solution using your rule from part **a**? Using your graph? Why or why not?

8. a. Use your graph to figure out how many hours Tonya must work to make $86.50. Explain your method.

   b. Check your solution to part **a** by using your equation. Show your work.

Writing

9. Answer the letter to Dr. Math.

> Dear Dr. Math,
> An advertisement said I could rent a video camera for a fee of $12 plus $5 per hour. I have $42 and am trying to figure out how long I can rent that camera. Can you help?
> Future Movie Director

Solving Simple Equations and Inequalities

Applying Skills

Find the value of x that solves each equation. Check your solution by substituting it back into the equation.

1. $4x + 3 = 11$ **2.** $4 + 2x = 10$

3. $3x + 5 = 26$ **4.** $-1 + 2x = 19$

5. $1.5x + 3 = 18$ **6.** $-5 + 7x = 30$

7. $4x + 7 = 35$ **8.** $-6x = 18$

9. $-2x + 3 = 7$ **10.** $5x + 8 = 38$

11. $2 + 3x = -10$ **12.** $-3x + 6 = 15$

Find the values of x that make each inequality true. Graph each solution on a number line.

13. $3x < 9$ **14.** $x + 2 < 5$

15. $x - 3 \geq 1$ **16.** $4x > 8$

17. $x - 2 > 0$ **18.** $2x + 4 \leq 8$

19. $x + 3 \leq 1$ **20.** $2x \geq 6$

21. $3x - 1 \geq 2$

Extending Concepts

Write each equation or inequality using symbols. Then solve the equation or inequality and explain how you solved it.

22. Ten more than x is 23.

23. Six times y plus 2 is 20.

24. Twice m is less than 6.

25. Two less than x is greater than 7.

Writing

26. Answer the letter to Dr. Math.

Dear Dr. Math,

My friend, Lee, and I were trying to solve the equation $4x + 2 = 21$. Lee began by saying, "Something plus 2 equals 21, so that 'something' must be equal to 19." Then he wrote $4x = 19$ and said, "4 times something is 19." I told him to stop right there—no number multiplied by 4 will give you 19. So, I think he made a mistake somewhere. Can you help us?

E. Quation

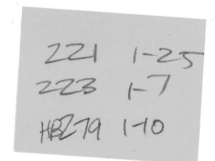

The Balancing Act

Applying Skills

Each picture shows a balanced scale. In each picture, all cups contain the same (unknown) number of marbles. Figure out how many marbles a cup contains.

1.

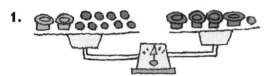

2.

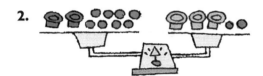

3.

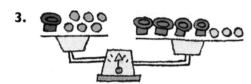

4.

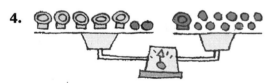

Solve each equation using any method. Keep a step-by-step record of your work and check your solution.

5. $3x + 8 = 4x + 1$

6. $2y + 8 = 5y + 2$

7. $3r + 13 = 7r + 1$

8. $5t + 7 = t + 9$

9. $x + 9 = 3x - 1$

10. $3x + 2 = 4x - 4$

Draw a balance picture for each equation and solve the equation.

11. $3m + 9 = 5m + 3$

12. $4x + 3 = 2x + 4$

13. $5y + 1 = 4y + 10$

Extending Concepts

Bernie's Bicycles rents bicycles for $10 plus $4 per hour.

14. You have $26. Write an equation that will give you the number of hours that you could rent a bike.

15. Draw a balance picture for the equation and then solve the equation.

Writing

16. Answer the letter to Dr. Math.

> Dear Dr. Math,
> I solved the inequality $2x \geq 3$ by testing lots of different values for x. I found that if x is 2, 3, 4, 5 or any larger number, the inequality is true. If x is 1, 0, or a negative number, it is false. So I figured that the solution to the inequality must be $x \geq 2$. Was this a good way to solve the inequality? What do you think?
> Jean Luc

The Best Bus Deal

Applying Skills

Sasha is trying to decide between two summer jobs. The first job is working as a baby-sitter for $6.00 per hour. The other job is a T-shirt vendor at the state park. This job pays $4.00 per hour plus $0.25 per T-shirt she sells.

1. If she works 23 hours as a baby-sitter, how much money will she earn?

2. If she works 23 hours at the state park and sells 13 T-shirts, how much money will she earn?

3. Write an equation that relates the number of hours worked *h* to the dollars earned *d* working as a baby-sitter.

4. Sasha assumes that she can sell at least 10 T-shirts per hour. Write an equation that relates the number of hours selling T-shirts *t* to the dollars earned *d*.

5. If Sasha works 40 hours, how many T-shirts would she have to sell to earn at least as much as she earns working as a baby-sitter?

Extending Concepts

The graph shows the expense of renting a row boat.

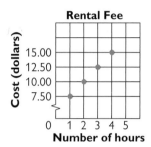

6. How much will it cost to rent a row boat for 5 hours?

7. If you have $20 to spend, for how many hours can you rent the boat?

Writing

8. Answer the letter to Dr. Math.

Dear Dr. Math,

When I solved the equation $2x + 1 = 17$, I drew a balance picture and wound up finding that $x = 8$. I checked my solution, so I know it's right, but someone told me I could have solved the equation a different way. What other ways could I have done it?

Sally D. Solver

Your first task is to analyze a simple business scenario. This will prepare you to play a simulation. In the simulation, you become the owner of a small food booth at a school fund-raising fair. Figuring out profit, income, and expenses is part of your responsibility as a business owner.

What math is used to increase profits?

GETTING DOWN TO BUSINESS

PHASE**TWO**
Spreadsheets

You will become familiar with using spreadsheets on the computer in this phase. Spreadsheet notation, formulas, and tips are all part of what you will learn about as you go along. Playing a game using a spreadsheet helps you get started. Toward the end of the phase, you will add data to a spreadsheet on the computer and use it to explore ways to make your food booth more profitable.

PHASE**THREE**
Spreadsheets and Graphs

Acting as a consultant to the Tee-Time T-Shirt Company, you will try to help Tee-Time find the best price to charge for T-shirts. Spreadsheets and graphs become your tools as you respond to Tee-Time's question. These tools will help you show the relationships between price and income, price and expenses, and price and profit.

PHASE**FOUR**
A Case Study of North Mall Cinema

Your final project is a case study of North Mall Cinema. You will play the role of an employee of Enterprise Consulting Company. As a consultant, you will use information from North Mall Cinema to calculate income and expenses. Your job is to help North Mall improve its profits. Building a spreadsheet from scratch on the computer will help you find ways to improve profit.

PHASE ONE

In this phase, you will run a food booth simulation. Figuring out profit, income, and expenses will help your simulation run smoothly.

Learning the relationships among profit, income, and expenses is an important skill for running a business, big or small. Why do you think this would be important?

Business Simulation

WHAT'S THE MATH?

Investigations in this section focus on:

DATA and STATISTICS

- Recording information in a table

ALGEBRA and FUNCTIONS

- Exploring the concepts of profit, income, and expenses

- Exploring the relationships among profit, income, and expenses

- Posing *what-if* questions that bring out functional relationships

- Calculating profit and income

MathScape Online
mathscape2.com/self_check_quiz

1 To Sell or Not to Sell Gourmet Hot Dogs

Imagine what it would be like to run your own business.
Businesses are always looking for ways to improve their profit. You will learn about the business terms *profit, income,* and *expenses* as you analyze this example of a food booth run by students.

Read the Hot Dog Stand Script

What are the meanings of profit, income, and expenses?

Read the information in the box The Hot Dog Stand Introduction. Then think about the meanings of profit, income, and expenses in business terms as you read the script on the handout The Hot Dog Stand Part 1.

The Hot Dog Stand Introduction

Scene: It is 8:00 AM at your favorite Middle School. Three students are setting up their hot dog stand as one of the food booths at the school fund-raising fair.

Cast: Student 1 is eager to run a hot dog stand, but wants it to be out of the ordinary. Student 2 agrees with Student 1. Student 3 wants to have a regular hot dog stand because people will be familiar with it.

Analyze the Hot Dog Stand Business

Talk with the members of your group about the Hot Dog Stand. Consider these questions in your investigation.

1 Do you think the Hot Dog Stand will make any profit? Why or why not?

2 Where will this business get its income? What are this business's expenses? Show your thinking.

3 What would you do differently if you were running the business?

4 Is there a chance they could lose money? How could that happen?

What do you know about profit, income, and expenses, and how they are related?

Improve the Hot Dog Stand

Write a letter to the students who are running the Hot Dog Stand, describing your suggestions for improving their business. Think about the decisions the students made when setting up the Hot Dog Stand. Include answers to these questions in your suggestions.

- What are the decisions the students made about setting up their business?

- Do you agree or disagree with each decision the students made? Why?

hot **words** | income
expense

HW**omework**

page 258

2 A Food Booth at a School Fair

How much money could you make from a food booth at the school fair? One way to find out is by setting up a simulation. Here you will play a simulation of a food booth. Think about how profit, income, and expenses are related, and how the decisions you make affect profit.

Play the Simulation with the Class

What are the relationships among income, expenses, and profit?

To play this simulation, you will need the following: Setting Up the Food Booth Simulation, Simulation Recording Sheet, Income Results, and Expense Results. Before the class begins playing, complete Setting Up the Food Booth Simulation.

As you play the simulation, think about what decisions you are making as owner of your food booth.

Simulation Rules

1. For each turn, a player rolls two number cubes. The player selects, reads aloud, and crosses off an event from Income Results. Each player records the information on the Simulation Recording Sheet.

2. The player rolls again to select, read aloud, and cross off an event from Expense Results. Each player again records the information on the Simulation Recording Sheet and calculates the profit for the turn. This ends one turn.

3. Players play for six turns, taking turns rolling cubes on each turn. If players roll a number they have already used, they roll again until they get a new number.

Play the Simulation in Pairs

Each pair will need two number cubes and two new Simulation Recording Sheets to play the simulation. To play with a partner, follow the same rules as you did when the class played the simulation.

Compare the Results of the Two Simulations

Think about the results of your simulation the second time you played. Describe how your profit changed.

- What decisions did you make the second time you played the simulation? Be sure to tell about the price you chose, how much inventory you bought, what choices you made from the numbers rolled, and any other decisions you made.

- Which of these choices did you make differently the second time?

- How did your final profit compare to your profit the first time you played the simulation?

- Which of your decisions do you think made the biggest difference in your profit the second time? Why?

How do your business decisions affect your profit?

hot words | profit simulation

Homework
page 259

3 What-If Questions for the Food Booth

What will happen to profit if you charge a different price at your food booth? You can explore different ways of running your food booth simulation by testing how changing price affects profit. See how well you can predict the effects of a price change on income and expenses.

Analyze a Business Decision

What if you changed the price at your food booth?

As you read the script on the handout The Hot Dog Stand Part 2, think about what might have happened if the Hot Dog Stand had been run differently. Choose a different price from the one you used in the simulation. Use the same information that you used in the simulation for how many you sold and how many you bought. Then figure out your profit.

Create What-If Questions

Use the Simulation Recording Sheet from Lesson 2 to respond to the following:

- List three what-if questions about your food booth.

- For each question, write about whether you think income and expenses will increase or decrease. Be sure to explain your thinking.

How could you pose what-if questions about your own booth?

Report on Profit, Income, and Expenses

Write a short report about how your food booth did at the fair. Your suggestions will be used to plan next year's fair. Include the following in your report:

- a brief description of what happened at your food booth

- a list of your income (Be sure to show how you calculated it.)

- a list of your expenses (Be sure to show how you calculated it.)

- your profit (Be sure to show how you calculated it.)

- three suggestions for making a better profit (Be sure to tell why they will result in a better profit.)

hot words | price
what-if questions

Homework
page 260

PHASE TWO

In this information age, you need to know how to use computers to be in business. You will learn in this phase what spreadsheets are and how to use them to enhance your business.

You will be introduced to the various elements of a spreadsheet. How is data arranged on a spreadsheet? How do columns and rows help someone read a spreadsheet?

Spreadsheets

WHAT'S THE MATH?

Investigations in this section focus on:

DATA and STATISTICS

- Recording and organizing information in a table
- Interpreting and analyzing information in a table

ALGEBRA and FUNCTIONS

- Posing what-if questions that bring out functional relationships
- Reading and writing formulas in spreadsheet notation
- Applying your knowledge of the order of operations to spreadsheet calculations

MathScape Online
mathscape2.com/self_check_quiz

4 What-If Questions on Spreadsheets

You can use a computer spreadsheet that has been set up with certain formulas to calculate profit, income, and expenses. In this lesson, you will explore what-if questions on a spreadsheet. This will help you to see how business decisions affect profit.

Investigate Profit on a Spreadsheet

How can you use a spreadsheet?

A **spreadsheet** is organized into a grid of rows and columns. **Rows** are horizontal and numbered along the left side. **Columns** are vertical and labeled across the top of the spreadsheet with letters. Each small rectangle is called a **cell.** Cells are where you place information or data in your spreadsheet. Look over the spreadsheet containing data on this page and answer these questions:

1 Where would you find all the information about popcorn? potato chips?

2 Where would you find all the information for expenses? for price? for income?

3 Where would you find the total profit? the number of chocolate chip cookies sold?

	A	B	C	D	E	F	G
	Booth	No. Sold	Price	Income	Cost per Item	Expenses	Profit
1							
2	Booth	No. Sold	Price	Income	Cost per Item	Expenses	Profit
3	Potato Chips	600	$0.30	$180.00	$0.15	$90.00	$90.00
4	Chocolate Chip Cookies	750	$0.40	$300.00	$0.15	$112.50	$187.50
5	Popcorn	800	$0.30	$240.00	$0.15	$120.00	$120.00
6	TOTALS						$397.50

Profit Made by the Food Booths

Explore What-If Questions on a Spreadsheet

How can you use a spreadsheet to explore what-if questions?

Use the spreadsheet on the computer to change the numbers described in each what-if question below.

1 What if each booth sells double the amount shown on the spreadsheet? Record the information in the Number Sold and Profit columns. Record the Total Profit.

2 What if each booth sells half as many items? Record the information in the Number Sold and Profit columns. Record the Total Profit.

3 What if the expenses for each booth double? Record the information in the Expenses and Profit columns. Record the Total Profit.

4 What if nobody buys potato chips and you bought 600 bags of chips? Record the information in the Number Sold and Profit columns. Record the Total Profit.

5 What if you raised your prices by 10 cents? Record the information in the Price and Profit columns. Record the Total Profit.

6 What if you raised your prices by 15 cents and your cost per item went up by 15 cents? Record the information in the Price, Expenses, and Profit columns. Record the Total Profit.

Tips for Using a Spreadsheet

- Find out the name of the file for your spreadsheet and open it.

- To change information in a cell, click the cell to select it. Type your changes into the formula bar and press RETURN or ENTER.

- Remember to save the file before you close the spreadsheet, so that you don't lose your data.

hot **words** | spreadsheet cells

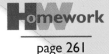

page 261

5 "What's My Formula?" Game

The secret to the power of a spreadsheet is in writing the formulas that make the calculations for you. Here you will play a game on the computer that will help you learn how to write formulas in spreadsheet notation. Can you figure out your partner's hidden formula?

Write Formulas in Spreadsheet Notation

How do you write formulas for a spreadsheet?

Read the guidelines below. Use the Profit Made by the Food Booths on page 236 to understand formulas in spreadsheets and answer these questions:

1 What cell would have the formula =B3*C3 in it? What information in the spreadsheet does this formula give you?

2 What cell would have the formula =D4−F4 in it? What information in the spreadsheet does this formula give you?

3 Write a spreadsheet formula that could make the calculation for each of the following: income for the popcorn booth, profit for the potato chips booth, and total profit.

Guidelines for Writing Spreadsheet Formulas

- Each cell has a letter and a number. For example, the cell in the upper left corner is named A1.

- In a spreadsheet, you use an asterisk (*) to multiply. For example, 2*3.

- In a spreadsheet, you use a forward slash (/) to divide. For example, 16/4.

- Formulas in spreadsheet notation usually begin with an = sign. (Some may use a + sign instead.) An example is =B4−7. One way to think about this is that you select a cell and then press = to tell the computer, "Make cell A1 equal to the answer to this equation."

Play a Formula Game

You and your partner will need a computer with a spreadsheet to play this game.

Can you figure out your partner's hidden formula?

Rules for "What's My Formula?"

Player A will make formulas, and Player B will guess hidden formulas. Decide who will be Players A and B. Players switch roles after several rounds.

1. Player B's eyes are closed while Player A enters a number into a cell.

	A	B	C
1			
2		6	
3			
4			

2. In any other cell, Player A enters a formula that uses the cell of the number just entered.

	A	B	C
1			
2		6	
3			
4			=B2+7

3. Player A presses RETURN or clicks the box. The formula's result will appear on the spreadsheet.

	A	B	C
1			
2		6	
3			
4			13

4. Player B tries to discover the formula. Player B can only change the number in the first cell where Player A entered a number.

	A	B	C
1			
2		→6	
3			
4			13

5. Player B checks an idea by entering it as a formula in a different cell and comparing answers. If they are the same, the formula is correct. Player B prints the spreadsheet.

	A	B	C
1			
2		6	
3			
4		=B2+7	13

Share About Spreadsheet Formulas

Write several tips for someone using a spreadsheet for the first time. Consider these questions:

- What did you learn about a spreadsheet that you didn't know before?

- What was confusing to you? What suggestions do you have for making it less confusing?

- What other helpful tips do you have about using a spreadsheet?

hot **words** | formula cells

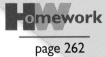

Homework

page 262

6 Double Your Profits?

CREATING A SPREADSHEET TO DOUBLE PROFITS

How could you double your profits for your food booth?

You can use what you have learned about spreadsheets to explore ways to increase your profits. Organizing information, posing what-if questions, and analyzing data will help you find ways to double profits.

Double Profits on a Spreadsheet

How could you double your profit for your food booth?

After you have planned your spreadsheet on paper, enter it into the computer. Use the spreadsheet you create on the computer to find three different ways you could double the profit of your simulation. For each way that you find, complete the following tasks:

- Write about it in the form of a what-if question.

- Keep track of what happens to income, expenses, and profit.

- Print the spreadsheet that shows the results.

Steps for Planning a Spreadsheet

1. Make two tables with the following column headings: Booth, Number Sold, Price, Income, Expenses, and Profit.

2. In the first table, enter the data from one of the turns on your Simulation Recording Sheet. Make sure you enter the food item you chose in the booth column.

3. Figure out the appropriate formulas using spreadsheet notation for income, expenses, and profit.

4. In the second table, enter these formulas in the correct cells.

Analyze Your Business

Use the information you gathered on the computer about doubling profits to answer these questions:

1 Describe the three ways you doubled your profit.

2 As a business owner, which way would you recommend for doubling your profit? Why?

Analyze Your Spreadsheet

Spend a few minutes reflecting on what you have learned about spreadsheets and formulas.

1 Use these questions to help you write about how you set up your spreadsheet:

a. What columns and rows did you put on your spreadsheet?

b. What formulas did you put into your spreadsheet?

c. How did you decide which cells should have formulas?

d. How would you explain to someone what a formula is?

2 Write three different what-if questions you could explore on your spreadsheet.

3 How would someone else test one of the what-if questions on your spreadsheet?

4 Attach a copy of your spreadsheet to your work.

What would you recommend for your business?

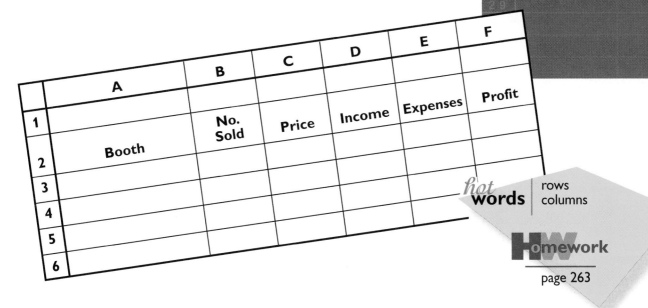

	A	B	C	D	E	F
1		No. Sold	Price	Income	Expenses	Profit
2	Booth					
3						
4						
5						
6						

hot **words** | rows columns

Homework
page 263

PHASE THREE

A

PRICE
$1200
$1300
$140
$150
$160

5

6

7

Graphs are helpful in showing relationships and trends. They are important tools for most businesses. In this phase, you will create graphs and spreadsheets that will help you recommend the best price for the Tee-Time T-Shirt Company to charge for T-shirts.

Think about graphs you might see in the business section of a newspaper, for example. How might those graphs show trends in profit, income, or expenses for a business?

Spreadsheets and Graphs

WHAT'S THE MATH?

Investigations in this section focus on:

DATA and STATISTICS

- Organizing data in a table
- Plotting data points on a graph
- Exploring relationships among points on a graph
- Creating and interpreting qualitative graphs

ALGEBRA and FUNCTIONS

- Reading and writing formulas in spreadsheet notation
- Analyzing spreadsheets and graphs to explore the relationships between them

MathScape Online
mathscape2.com/self_check_quiz

7 How Many Sales at Tee-Time?

Tee-Time T-Shirt Company wants help in deciding what price customers would be willing to pay for its T-shirts.
A survey that your class conducts will provide data you can graph. You will explore the relationship between price and the number of people who would buy T-shirts.

Predict How Price Will Affect Sales

How might the price of T-shirts affect sales?

Does charging more for a product always mean you make more money? Think about the Tee-Time T-Shirt Company presentation. Write a prediction about how the price that Tee-Time charges might affect sales.

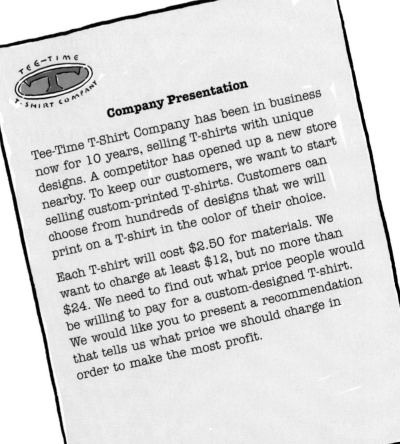

Company Presentation

Tee-Time T-Shirt Company has been in business now for 10 years, selling T-shirts with unique designs. A competitor has opened up a new store nearby. To keep our customers, we want to start selling custom-printed T-shirts. Customers can choose from hundreds of designs that we will print on a T-shirt in the color of their choice.

Each T-shirt will cost $2.50 for materials. We want to charge at least $12, but no more than $24. We need to find out what price people would be willing to pay for a custom-designed T-shirt. We would like you to present a recommendation that tells us what price we should charge in order to make the most profit.

Plot Class Data on a Graph

After your class has conducted a survey on what price people would pay for a T-shirt, set up your own graph showing the class data.

- Set up the graph to show the number of students who said they would buy a T-shirt at each price listed in the survey.

- Plot the points on your graph. Do not draw lines that connect the points.

- Give your graph a title.

How can you make a graph that shows the price people would pay for the product?

Analyze Your Graph

Think about how you could use your graph to explore the relationship between price and the number of people who will buy the product. Answer these questions.

- What trend does the graph show about sales and price?

- Does this correspond to what you know about the number of people who buy an item when its price increases? Explain your thinking.

- What do you think could happen to profit as the price increases?

hot **words** | price
scatter plot

Homework
page 264

8 How Much Profit at Tee-Time?

RELATING GRAPHS
AND SPREADSHEETS

What do you think is the best price for Tee-Time to charge for T-shirts? As you continue to investigate this question, you will explore spreadsheets and their related graphs. This will help you to better understand the relationships among price, income, and expenses.

Use a Spreadsheet to Find the Best Price

How can you use your spreadsheet to find a closer estimate for the best price?

Create your spreadsheet on the computer. Use it to test out different prices to find the highest profit. Make sure you do the following:

1 Keep track of the different prices you try, the number of sales, and the amount of profit at each price.

2 When you have decided on the best price to charge, write your recommendation letter to Tee-Time. Include a clear explanation of the price that will result in the most profit and how your information led you to make this recommendation.

3 Make a copy of your spreadsheet with the best price highlighted. Attach it to your letter.

Information to Include in Your Spreadsheet

- different possible prices (from the class survey)

- number of sales at each different price (from the class survey)

- income at each price

- expenses (Base this on 40 shirts at a cost to us of $2.50 per shirt. Figure on $100 for expenses at whatever price we charge.)

- the profit at each price

Make Graphs from Spreadsheets

Graphs will help Tee-Time see how profit, income, and expenses are related. Use the data from your spreadsheet to plot points on three separate graphs:

1 Create a graph showing price versus income.

2 Create a graph showing price versus expenses.

3 Create a graph showing price versus profit.

Attach these three graphs to your recommendation letter and spreadsheet.

How can you show the relationships among price, income, and expenses on a graph?

Analyze Your Graphs

To analyze the relationship between price and profit, use the three graphs that you made to write answers to these questions:

- How would you describe the shape formed by the points in each of your three graphs? Do the graphs form straight lines? Do they change directions? At what points? Why?

- What does the shape of the graph tell you about the relationship between price and income? price and expenses? price and profit?

- If you compare all three graphs, what can you learn about income, expenses, and profit?

- Does the profit graph ever fall below zero? At what point? Why?

- From the information on the graph, what do you think would be the most profitable price for Tee-Time to charge? Why?

hot **words** | spreadsheet profit

Homework

page 265

9 Months Later at Tee-Time

CREATING AND
INTERPRETING
QUALITATIVE
GRAPHS

Businesses often try to predict trends that will affect their profits. Graphs that have no numbers can be a useful way of showing general trends. As you create and interpret these kinds of graphs, you will see relationships among profit, income, and expenses in a new way.

Make Qualitative Graphs

How could you make a graph without numbers?

Tee-Time T-Shirt Company wants your help in preparing graphs to present to their board of directors. They don't want the board to look at lots of numbers. They want these graphs to show general trends in the business over the past 6 months. Use the information in the Tee-Time T-Shirt Company Semi-Annual Report to make a graph of their profit without numbers.

Semi-Annual Report

Aug. 1 We introduce our new custom designed T-shirts. We are the only T-shirt company selling them, so our sales increase quite a lot. Our expenses rise only a little. Sales keep increasing until Sept. 15.

Sept. 16 We decide to hire a local celebrity, Arnold Quartzdigger, to do our TV commercials for us. Mr. Quartzdigger asks for a high salary, so our expenses increase quite a bit. On Oct. 2, we ask him to leave because he is too expensive.

Oct. 3 We learn that our competitor, Weekend Wear Clothing, has started selling custom designed T-shirts, and is selling them for less. We find a supplier that will sell us T-shirts and designs for less than we were paying. We lower our price below Weekend Wear's. Our income drops a little, but our expenses also drop by about the same amount.

Nov. 15 We have expanded our line of custom designed T-shirts to include six new colors that Weekend Wear does not have. Our sales increase gradually until Dec. 1 while our expenses stay the same.

Dec. 1 Many people are buying gifts for the holidays, so we lower our prices slightly to encourage more people to buy from us. It works! Our sales keep going up throughout the month of December, while our expenses stay about the same.

Jan. 5 The holidays are over, and people are not spending as much money. This month we don't sell many T-shirts, even though our expenses remain about the same.

Create the Missing Graphs

The handout Missing Graphs is a collection of graphs showing general trends for the Tee-Time T-Shirt Company. The graphs show increases and decreases in income, expenses, and profit during the last seven months. Use the information from each pair of graphs to sketch the missing third graph.

Given two out of three graphs, how could you create the third graph?

Describe Income, Expenses, and Profit

Read the memo below from the manager of Enterprise Consultants, Inc. Provide the information the manager needs for the presentation to Tee-Time T-Shirt Company.

To: Employees of Enterprise Consultants, Inc.

You have done an excellent job interpreting the information from Tee-Time. I am now preparing to give a presentation to Tee-Time to teach them more about profit, income, and expenses. I would like to have some posters to help me explain concepts clearly to Tee-Time's employees.

Here is what I'd like you to do:

1. Create a poster with two or three sets of profit, income, and expense graphs.

2. Explain in writing what each graph shows.

3. Explain in writing how the income and expense graphs result in the profit graph.

hot **words** | qualitative graphs
trend

omework
page 266

PHASE FOUR

PROD. NO.
SCENE
DATE
PROD. CO.
DIRECTOR

Your final project runs through this entire phase and gives you the chance to play the role of a consultant to North Mall Cinema. Your job is to recommend what they can do to increase profits. What you have learned in this unit will come into play—spreadsheets, what-if questions, formulas, profit, income, expenses, and more!

Try to imagine what it would be like to run a cinema. What kinds of things would you need to know to keep the business profitable?

A Case Study of North Mall Cinema

WHAT'S THE MATH?

Investigations in this section focus on:

DATA and STATISTICS

- Recording and organizing information in a table

- Interpreting and analyzing information in a table

ALGEBRA and FUNCTIONS

- Calculating profit and income

- Posing what-if questions that bring out the relationships among profit, income, and expenses

MathScape Online
mathscape2.com/self_check_quiz

10

North Mall Cinema's Project

As a final project, you will act as a consultant to a business. Your job will be to research how North Mall Cinema can increase their profits. The first two steps in the project are organizing information about the business and listing what-if questions to explore.

Organize Information About North Mall Cinema

How could you determine the income and expenses for each theater per day?

Read the memo from North Mall Cinema outlining your final project and the second part of the handout North Mall Cinema, entitled Information About North Mall Cinema. To complete Step 1 of your final project, do this:

- Make a list of income and expenses for each theater per day.

- Calculate income and expenses for each theater per day. Show your work.

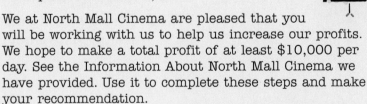

To: Consultants of
Enterprise Consultants, Inc.

We at North Mall Cinema are pleased that you will be working with us to help us increase our profits. We hope to make a total profit of at least $10,000 per day. See the Information About North Mall Cinema we have provided. Use it to complete these steps and make your recommendation.

Step 1. Organize data provided about North Mall Cinema.

Step 2. List what-if questions to explore.

Step 3. Design a spreadsheet.

Step 4. Investigate profit on the spreadsheet.

Step 5. Create a report on improving profit.

List What-If Questions to Explore

Think of suggestions that might make North Mall Cinema more profitable before you complete the tasks below. Brainstorming ideas with a partner might be helpful.

1 Use the suggestions to make a list of what-if questions to explore on the spreadsheet. List at least seven what-if questions.

2 From your list of what-if questions, select three questions that you feel would have the greatest impact on North Mall Cinema's profits and write them down.

How can North Mall Cinema increase its profits, and what questions would you ask to find out?

hot **words** | profit
what-if questions

HW**omework**

page 267

11 North Mall Cinema's Spreadsheet

What spreadsheet can you create to help you decide how to increase North Mall Cinema's profits? To complete the third and fourth steps of the project, you will design a spreadsheet on the computer and use it to explore the what-if questions you came up with in the last lesson.

How could you design a spreadsheet showing information about North Mall Cinema?

Design a Spreadsheet

Design your own spreadsheet. Use the second part of the handout North Mall Cinema, entitled Information About North Mall Cinema. Create the spreadsheet on paper first. You will be using the spreadsheet you create on paper to build a spreadsheet on the computer later on.

- Be sure your spreadsheet calculates the total income, the total expenses, and the profit for each of the four theaters at the North Mall Cinema.

- As you design your spreadsheet, you might want to discuss ideas with your partner.

Investigate Profit on the Spreadsheet

Using the spreadsheet you designed on paper, build your spreadsheet on the computer. Then use it to explore each of the three what-if questions you wrote in the last lesson. For each question you explore, write the following information:

- the question you are exploring

- what you tried on the spreadsheet

- whether it helped you reach North Mall Cinema's goal of making at least $10,000 profit each day

- whether you increased or decreased income or expenses

How can you use a spreadsheet to explore your what-if questions?

	A	B	C	D	E	F
1						
2						
3						
4						
5						
6						
7						
8						
9						
10						
11						
12						
13						
14						
15						
16						
17						
18						
19						
20						
21						
22						
23						
24						
25						
26						
27						
28						
29						
30						
31						
32						
33						
34						

hot **words** | spreadsheet
formula

Homework

page 268

12 North Mall Cinema's Report

MAKING PROFIT RECOMMENDATIONS

The feedback of a partner can be helpful in refining your work. After writing a report for your final project, you will have a chance to revise it and make improvements based on your partner's feedback. You will want your report to look as professional as possible.

How would you write a report that gives recommendations for improving profit?

Create a Report on Improving Profit

Write a report to North Mall Cinema. Include the following in your report:

- a description of one, two, or three recommendations for meeting North Mall Cinema's profit goal of at least $10,000 per day

- an explanation of each recommendation

- how your recommendation(s) are related to profit, income, and expenses

Attach to your report the what-if questions you explored, the spreadsheet you designed, and the information you got from your spreadsheet.

Provide Feedback on a Partner's Report

Read your partner's report carefully and write some ways it could be improved. Here are suggestions to help you get started:

- I really like how you did . . .

- Some things you could improve are . . .

- I had trouble understanding what you meant by . . .

- The part that seemed unclear to me was when you said . . .

What suggestions do you have for helping your partner improve his or her work?

Revise the Report

Read your partner's feedback on your report. Think about your partner's suggestions and your own ideas for improving your report. Then revise your report. When you have finished revising your report, pull together these pieces of your final project:

- your revised report

- your original report

- your spreadsheet design on paper

- your spreadsheet printout

- your what-if questions

hot **words** | income expense

Homework

page 269

 **Homework 1**

To Sell or Not to Sell Gourmet Hot Dogs

Applying Skills

1. Make two columns. Label one column "Expenses" and the other "Income." Put the following business situations for Newburger's Seasoned Hot Dog Booth under the appropriate column.

   a. Paid for photocopying expenses to make flyers for advertising.

   b. Purchased food from Harry and Don's Hot Dog Distributing Company.

   c. Sold 20 hot dogs to a family for a cookout.

   d. Purchased relish, ketchup, and mustard from a supermarket.

2. As owner of Anne's Amorphous Shoe Store, decide which option listed below (A, B, or C) you will choose to advertise your upcoming shoe sale and explain why. Use the information in the table shown to make your decision. You will need to call your local copying center and newspaper to get more information.

Option	Type of Advertising	How Many People Will Know About Sale	Cost
A	100 flyers	500	call for photocopying cost
B	half-page newspaper advertisement	1,200	call for newspaper advertising costs
C	sign in window	200	$5.00

Extending Concepts

3. If you were the owner of Newburger's Seasoned Hot Dog Booth, what information would you need to decide how much you should charge the customer for each hot dog?

4. How would you know if your hot dog booth was successful?

5. If Newburger's Seasoned Hot Dog Booth's income increased by $10 and its expenses remained the same, what would happen to the hot dog booth's profit?

Making Connections

The relationship between income, expenses, and profit is similar to the relationship between deposits, withdrawals, and account balance. Example: I have $300 in the bank and deposit $200. If I withdraw $50, my account balance would be $450.

Transaction Statement for: _____
Account #: 453565233

Date	Transaction Type	Amount	Account Balance
04/27/05	Open Account	$300.00	$300.00
04/28/05	Deposit	$200.00	$500.00
05/01/05	Withdrawal	$50.00	$450.00

6. Explain how deposit, withdrawal, and account balance are related to income, expenses, and profit.

7. Calculate the final account balance after examining the Transaction Statement for Account #7896543.

Transaction Statement for: _____
Account #: 7896543

Account balance before first transaction on this statement: $525.00

Date	Transaction Type	Amount
06/01/05	Withdrawal	$234.65
06/03/05	Withdrawal	$92.75
06/12/05	Deposit	$253.92
06/12/05	Withdrawal	$210.35
06/18/05	Withdrawal	$15.75

A Food Booth at a School Fair

Applying Skills

Last April, it cost Newburger's Seasoned Hot Dog Booth $50 to buy a dozen jumbo packages of hot dogs and rolls from Harry and Don's Hot Dog Distributing Company. Each jumbo package contains 10 hot dogs and 10 rolls.

1. As the manager of Newburger's Seasoned Hot Dog Booth, you would like to find out how profit increases when the price of each hot dog increases. Use the information above to help you make a table as shown and fill it in.

Cost of Hot Dog on a Roll	Price of Hot Dog on a Roll	Profit from Selling 1 Hot Dog on a Roll
	$0.50	
	$0.75	
	$1.00	
	$1.25	

2. How much should you sell each hot dog for if you would like to break even?

3. If it now costs $75 to buy a dozen jumbo packages of hot dogs and rolls, how much would you sell the hot dogs for if you would like Newburger's Seasoned Hot Dog Booth to break even?

Unfortunately, it rained during the Fall Fair this year so not as many people showed up at Newburger's Seasoned Hot Dog Booth. As a result, the hot dog booth sold only 300 hot dogs instead of the expected 600 hot dogs. The owner, Mr. Newburger, bought 600 hot dogs for $0.55 each and sold them for $2.00 each.

4. How much profit would Mr. Newburger have made if he had sold 600 hot dogs?

5. How much profit did Mr. Newburger make selling 300 hot dogs?

6. How much expected profit did Mr. Newburger lose because of the rain?

Extending Concepts

7. In general, if income decreases by half and expenses remain the same, does that mean that profit will decrease by half?

8. Give an example of a situation where income is cut in half, expenses are cut in half, and profit is cut in half. Include numbers in your example.

9. One way to calculate the profit from selling 100 soccer balls is to subtract the total expenses of buying the soccer balls from the total income from selling the soccer balls. If you subtract the cost of buying one soccer ball from the price of selling one soccer ball and then multiply by 100, is this another way to calculate the correct profit? If so, why does this work? Will it always work?

Making Connections

10. Use the following table to decide from which bank you would like to obtain a one-year, $3,000 loan. Explain why. Support your decision with calculations. If you are not sure about how to calculate interest, ask your teacher or parent.

Name of Bank	Length of Loan	Interest Rate	Service Fee
Alligator Bank	1 year	2.9%	$200
Barracuda Bank	1 year	3.2%	none

What-If Questions for the Food Booth

Applying Skills

Imagine that you are a consultant to Bernie's Burger Restaurant and have the following information:

Bernie's Burger Restaurant

Income and Expense Report

INCOME $250 per week
EXPENSES $95 per week

1. What is Bernie's Burger Restaurant's profit per week?

2. If you give the restaurant advice that will double its income, how much more profit will it make per week?

3. If the restaurant's income is doubled, how much more profit will they make in a year?

4. Name two ways income can be increased.

5. If expenses increase $30 per week, how much does income need to increase so that profit does not change?

A local businesswoman, Ann Tagoni, opened a hot dog booth to compete with Newburger's Seasoned Hot Dog Booth. Ms. Tagoni's records for one day of selling hot dogs at the Fall Fair are shown in the table below.

Type of Hot Dog	Number Bought	Cost	Number Sold	Price
Plain	200	$0.15	176	$1.35
Gourmet	100	$0.55	80	$2.10

6. Make a table like the one shown below and fill it in using the information from the Table of Ms. Tagoni's Records.

Type of Hot Dog	Income	Expenses	Profit
Plain			
Gourmet			

7. If Ann Tagoni wanted to make the most profit, should she sell plain or gourmet hot dogs, based on her Fall Fair records?

Extending Concepts

8. Explain two changes in the entries in the Table of Ms. Tagoni's Records that she could make so that the profit from selling plain and gourmet hot dogs would be equal.

9. If you were the person in charge of buying hot dogs for Ann Tagoni's hot dog stand, how would you determine how many hot dogs to buy so that you do not buy too many?

10. Which is more profitable: overestimating by 24 plain hot dogs like Ms. Tagoni did, or underestimating by 24 plain hot dogs and buying only 152, all of which sold?

Writing

11. One of Ann Tagoni's employees thinks that the price of gourmet hot dogs should be a lot higher. Write about what would happen if Ms. Tagoni raised the price of gourmet hot dogs really high and what effect that would have on profit.

What-If Questions on Spreadsheets

Applying Skills

Use the following information about Bodacious Bicycles for items **1–4**.

Information for Making the Spreadsheet

- The store sells four brands of bicycles at the following prices:

Trek	$225
KTS	$275
Bianchi	$285
Gary Fisher	$300

- This is a list of the bicycles from lowest to highest cost to the store: Trek, KTS, Bianchi, and Gary Fisher.

- It costs the store $110 to buy a Trek bicycle from the manufacturing company. The cost of each bike increases by $20.

- In 1996, the store sold 300 Treks, 150 KTS's, 125 Bianchis, and 80 Gary Fishers.

---

1. Make a spreadsheet with these column headings: Brand of Bicycle, Number Sold, Price, Income, Cost, Expenses, and Profit. Label the columns with letters and rows with numbers. Fill in the data in your spreadsheet using the information given.

2. What if the store sold 20 fewer Trek bicycles and increased the number of KTS bicycles sold by 20 bicycles? Tell what the total profit would be.

3. What if each of the Gary Fisher bicycles had been sold for $35 more than the price shown? Tell what the store's new total profit from the sales of all of its bicycles would be. Which cells did you add to find the answer?

4. What if the store owner wanted the total profit to be $501 more and he could only change the price of Trek bicycles? Figure out how much he would have to charge for Trek bicycles.

Extending Concepts

Use the spreadsheet shown of Linton's Lemon Shop's business for April to complete items **5–6**.

	A	B	C	D	E	F	G
1	Item	Number Sold	Price	Income	Cost per Item	Expenses	Profit
2							
3	Lemons	15,000	$0.18	$2,700.00	$0.05	$750.00	$1,950.00

5. If cost doubled, how many cells will change? Which cell(s) will change?

6. If cost doubled and price doubled, what will the new profit be?

Writing

7. Write at least three what-if questions for Bodacious Bicycles. Tell how each question would affect their profit, income, and expenses.

"What's My Formula?" Game

Applying Skills

A spreadsheet of Caryn's Creative Card Store's business for the month of March is shown.

	A	B	C	D	E	F	G
1	Type of Card	Number Sold	Price	Income	Cost per Item	Expenses	Profit
2	Blank	300	$1.65	$495.00	$0.75	$225.00	$270.00
3	Birthday	450	$2.15	$967.50	$0.85	$382.50	$585.00
4	Anniversary	125	$1.90	$237.50	$0.85	$106.25	$131.25
5	Thank you	120	$1.65	$198.00	$0.60	$72.00	$126.00
6	Get well	80	$1.40	$112.00	$0.55	$44.00	$68.00
7	Total						$1,180.25

1. What cell would have the formula $=B2*C2$?

2. What cell would have the formula $=D4-F4$?

3. What cell would have the formula $=B5*E5$?

4. Which cell has a spreadsheet formula that uses more than two different cells?

5. Write the formula for cell F2.

6. Write the formula for cell G5.

7. Write the formula for cell G7.

8. If the information in cell C4 changes, which cells will also change?

9. Write a spreadsheet formula that could make the calculation for the total expenses. In what new cell would you put this formula?

Extending Concepts

10. If you decided not to have an income column and expenses column, what would be the formula for cell E5 for Caryn's Creative Card Store spreadsheet? Assume that the remaining cells have been shifted to the left.

11. If you decided not to have a column for profit, but instead only a cell for total profit, what would be the formula for the Total Profit cell for Caryn's Creative Card Store spreadsheet?

12. Not everyone who uses spreadsheets knows how to write spreadsheet formulas. Mario just computed the expenses, income, and profit on a calculator, then entered them in the spreadsheet instead of using the spreadsheet formulas so the computer can do the computation. If you were looking at a spreadsheet on a computer, how would you be able to determine if the total profit was calculated by a calculator or a spreadsheet formula? (Hint: What if you could change only one cell?)

Making Connections

13. Explain how a teacher could use a spreadsheet to record students' grade information and how it can help save time when computing grade averages. Imagine that you are the teacher. Make a spreadsheet that you would set up which would keep track of information in the most efficient way using these column headings: Student Name, Quiz 1, Quiz 2, Quiz 3, Test 1, Test 2, Test 3, Quiz Average, and Test Average. Make up quiz and test data, and then add formulas to the spreadsheet.

Double Your Profits?

Applying Skills

A spreadsheet for a simulation of Tobias's Tire Company is shown.

1. If you wanted to double the profit of your simulation, which cells might have their values increased?

	A	B	C	D	E	F	G
1	Type of Tire	Number Sold	Price	Income	Cost per Item	Expenses	Profit
2	Regular	60	$60.95	$3,657.00	$25.00	$1,500.00	$2,157.00
3	Steel-belted	82	$82.95	$6,801.90	$47.00	$3,854.00	$2,947.90
4	Snow tires	20	$75.95	$1,519.00	$36.00	$720.00	$799.00
5							
6	Total						$5,903.90

2. If you wanted to double the profit of your simulation, which cells might have their values decreased?

3. If you wanted the profit to increase but the price of each tire to remain the same, which cells would show changed values?

4. Which cells should have formulas?

5. Suppose someone asked what would happen to the profit if the cost of one type of tire increased by $15. Choose one type of tire. Then draw a spreadsheet that shows how you changed the data.

Extending Concepts

6. How could Tobias's Tire Company double its profit, but keep its prices low and therefore keep its customers happy? Explain your answer.

7. If companies want to double profit, which do you think they have more control over, reducing cost or raising price? Explain your reasoning.

Making Connections

The cardiac cycle consists of the contraction and relaxation of the chambers of the heart: the atriums and ventricles.

With each beat of the heart, about 70 mL of blood is pumped into the aorta, the artery that carries blood throughout the body. At a heart rate of 75 beats per minute, 5,250 mL of blood is pumped each minute. This is called the *minute output*.

To find out the length of each cardiac cycle, a scientist divides 60 seconds by the heart rate.

8. Use the facts above to make a spreadsheet like the one shown and fill it in.

	A	B	C
1	Heart Rate	Minute Output (mL)	Length of each Cycle (sec)
2	75	5,250	0.8
3	60		
4	82		
5	65		
6	(your rate)		

9. Find your own pulse and use it to measure your heart rate (how many times your heart beats in 1 minute). Enter information about your heart rate in the three columns of the spreadsheet.

10. Write formulas for columns B and C.

How Many Sales at Tee-Time?

Applying Skills

Hank's Hardware makes and sells tools. Use the spreadsheet below to complete items **1–4.**

	A	B	C
	Potential Price per Hammer	Number of Hammers that Consumers Would Buy at This Price	Potential Income from Hammer Sales
1			
2	$9.00	240	
3	$12.00		$2,592.00
4		150	$2,550.00
5	$22.00	125	
6		98	$2,744.00

1. Make a spreadsheet like the one shown for Hank's Hardware and fill in the blank cells of the spreadsheet.

2. If you knew that the profit earned from selling the $12.00 hammers was $1,728.00, what was the cost of buying each hammer?

3. If the cost of hammers increased to $6.25, which price should the hammer be sold for to make the greatest profit?

Extending Concepts

4. Make a graph using the spreadsheet from Hank's Hardware. Graph the number of hammers that would be bought by consumers at this price vs. the potential price of the hammer.

Making Connections

Economists interpret graphs like the one you drew for Hank's Hardware in item **4** along with another graph to determine how much to charge for an item.

The graph you made for Hank's Hardware would be called the *demand curve* because those prices represent what the consumers are willing to pay. In economics, accompanying the demand curve is a supply curve. The *supply curve* is determined by the number of items the manufacturer is willing to produce at a certain price. The *intersection point* of these two curves usually is the price for which the items are sold.

5. The spreadsheet shown indicates the number of hammers Hank is willing to produce at certain prices. Graph the information from this spreadsheet on your graph from item **4.** Find the intersection point to determine how much a hammer will sell for.

	Price per Hammer	Number of Hammers the Manufacturer Is Willing to Produce at This Price
1		
2	$9.00	72
3	$12.00	105
4	$17.00	150
5	$22.00	185
6	$30.00	225

6. You are hired as a business consultant by Swish Internet Company, a new business that bills customers for time spent using the Internet. The company would like you to create a spreadsheet to keep track of the time people spend on the Internet. Include the following information in your spreadsheet:

   a. phone number of the modem being used

   b. beginning time and ending time of use

   c. date

   d. amount of time spent using the Internet

   Make up some data for your spreadsheet and then add formulas to your spreadsheet.

How Much Profit at Tee-Time?

Applying Skills

Caryn's Creative Card Store wants to project the sales of birthday cards in the future. Caryn figures that she will sell approximately 5,500 birthday cards per year. The cost of the cards will increase $0.05 per year so the price she will sell them for will increase by $0.10 per year. Use the data in this table to make graphs and answer questions.

	A	B	C	D	E	F
1	Year	Price of Birthday Cards ($)	Income ($)	Cost of Birthday Cards ($)	Expense ($)	Profit ($)
2	1996	2.15	11,825.00	0.85	4,675.00	7,150.00
3	1997	2.25	12,375.00	0.90	4,950.00	7,425.00
4	1998	2.35	12,925.00	0.95	5,225.00	7,700.00
5	1999	2.45	13,475.00	1.00	5,500.00	7,975.00
6	2000	2.55	14,025.00	1.05	5,775.00	8,250.00
7	2001	2.65	14,575.00	1.10	6,050.00	8,525.00

1. Plot a price vs. income graph.

2. Plot a price vs. expenses graph.

3. Plot a price vs. profit graph.

4. Does the price vs. income graph look similar to the price vs. profit graph? Why or why not?

Use the graphs shown to complete items 5–7.

5. What information does point **A** on the graph give you?

6. What information does point **B** on the graph give you?

7. What information does point **C** on the graph give you?

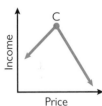

Use the data from Caryn's Creative Card Store for the years 1999–2004 to complete items 8–9.

8. Plot a cost vs. expenses graph.

9. Plot a cost vs. profit graph.

Extending Concepts

10. Why do your graphs for price vs. expenses and price vs. profit look similar?

11. The price vs. profit graph is different from the price vs. income graph. What do you think has caused this change?

Writing

12. Write a paragraph explaining how graphs and spreadsheets can be related to each other. Be sure to give examples.

Months Later at Tee-Time

Applying Skills

Draw a rough sketch of a profits vs. months graph for each description of these companies.

1. The Contemporary Video Store's March profits doubled from February. April was not as strong, especially since they had a lot of expenses for repairing their store after a severe storm. During the summer months the store righted itself and, by the end of August, its profits were almost the same as those earned in March. The profits remained steady until the end of the year. The company decided to keep prices the same throughout this entire period.

2. Anjali's Antique Store opened its doors in March. Tourist season was just beginning, so Anjali had a great number of customers stop by to buy antiques. Sales were steady throughout April, May, June, and July, but dropped in the beginning of August. Anjali's business really struggled in September, October, and November when it became cold. Sadly, Anjali had to close her store in December when she could only break even from selling antiques.

Read over the Action and Result columns for Tamika's Telecommunications Store, which sells fax machines. For items 3–7, write a brief description of why the result occurred at the store. There is more than one possibility.

	Action	Result
3.	Price of faxes decreases.	Profit increases.
4.	Cost of faxes increases.	Profit stays the same.
5.	Price of faxes decreases.	Profit stays the same.
6.	Phone company decreases the cost of using the phone.	Profit stays the same.
7.	Internet increases in popularity.	Profits decrease.

Extending Concepts

8. Identify the labels for each axis of this graph. Explain your labels using what you have learned about profit, income, and expenses.

Making Connections

The status of the Standard and Poor's 500 Index, or S&P 500 as it is commonly referred to, is a good measuring tool to see how the New York Stock Exchange is performing.

9. Look at the graph that shows the general trend of the S&P 500 Index and describe what has occurred in the past ten years.

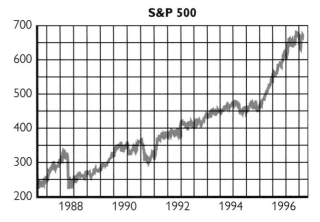

10. Make a prediction of how you think the S&P 500 Index will perform in the next ten years by sketching a graph like the one provided. Explain your reasoning.

North Mall Cinema's Project

Applying Skills

Linton's Lemon Shop

	A	B	C	D	E	F	G	H
	Month	**Item**	**Number Sold**	**Price**	**Income**	**Cost per Item**	**Expenses**	**Profit**
2	April	Lemons	15,000	$0.18			$750.00	$1,950.00
3	May	Lemons	13,200		$2,508.00	$0.06		

Marty and Margie's Lemon Shop

	A	B	C	D	E	F	G	H
	Month	**Item**	**Number Sold**	**Price**	**Income**	**Cost per Item**	**Expenses**	**Profit**
2	April	Lemons	8,000	$0.28			$800.00	$1,440.00
3	May	Lemons		$0.25	$2,995.00	$0.11		

Linton's Lemon Shop has a new competitor, Marty and Margie's Lemon Shop, which sells organically grown lemons. Organic lemons are more yellow and stay fresh longer, but they are more expensive. Above is a comparison of business for the months of April and May between Linton's Lemon Shop and Marty and Margie's Lemon Shop.

1. Make two spreadsheets as shown and fill in the blanks in the two spreadsheets.

2. Explain which shop is doing better and why.

Extending Concepts

The Bridge, a store specializing in children's ski clothes, has just opened.

3. Put the following situations in order, based on which you think has the greatest effect on The Bridge's profits: weather; hours store is open; cost of items in the store; price of items in the store; month; salary of employees; size of store; number of families living in the area; and how near the store is to the ski slopes. The situation which will have the greatest effect should be listed first, followed by the rest in descending order.

4. Describe the situations that The Bridge can and cannot control.

Making Connections

5. Using the information below, design a spreadsheet that would keep track of items, cost, and price, and calculate total profit. Include the data given and add formulas.

Item	Cost ($)	Price ($)
Soda	1.69	2.39
Shaving cream	1.15	2.19
Tomato sauce	2.09	4.99
Carrots	0.54	1.22
Toothpaste	0.89	1.76
Pasta	3.21	4.56
Beans	1.41	2.53
Rice	0.99	1.47

North Mall Cinema's Spreadsheet

Applying Skills

	A	B	C	D	E	F	G
1	Item	Number Sold	Price	Income	Cost per Item	Expenses	Profit
2	Tape	300	$1.45				$270.00
3	Box of staples		$1.85	$832.50			$585.00
4	Binder	125	$2.50		$1.45		
5	Box of paper clips	182		$318.50	$0.60		
6	Colored pencils		$3.40		$0.85		
7	Totals	1,131					

1. Copy the spreadsheet shown and fill in the blanks by using the clues the spreadsheet provides for you.

2. How would the total profit differ if 200 more colored pencils are sold?

3. If the cost of each box of staples increased by $0.20, what would the new price of a box of staples be in order for profit to remain the same?

4. If the price and cost of binders and tape doubled, how would total profit be affected?

Extending Concepts

5. Using these graphs, make a spreadsheet with these column headings: Year, Number Sold, Price, Income, Cost per Item, Expenses, and Profit. Fill in the data and formulas, and find the total profit.

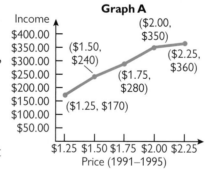

Graph A

Income
$400.00
$350.00 ($1.50, $240)
$300.00
$250.00
$200.00
$150.00 ($1.25, $170)
$100.00
$50.00

($2.00, $350)
($2.25, $360)
($1.75, $280)

$1.25 $1.50 $1.75 $2.00 $2.25
Price (1991–1995)

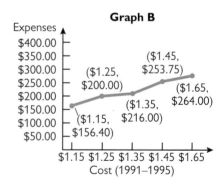

Graph B

Expenses
$400.00
$350.00
$300.00 ($1.25, $200.00)
$250.00
$200.00
$150.00 ($1.15, $156.40)
$100.00
$50.00

($1.45, $253.75)
($1.65, $264.00)
($1.35, $216.00)

$1.15 $1.25 $1.35 $1.45 $1.65
Cost (1991–1995)

6. Make a graph showing profit vs. year. Title it Graph C.

Writing

7. Write a paragraph describing what it would indicate about the relationship between year and profit as the years progressed if Graph A were steeper.

8. Describe the shape of Graph B in a sentence or two. What would have to occur for Graph B to be a straight line?

North Mall Cinema's Report

Applying Skills

The following spreadsheet gives data about the 12 languages spoken by the most people in the world. The spreadsheet identifies the 12 languages and divides the people who speak each language into two categories: native speakers and non-native speakers.

1. Copy the spreadsheet shown and fill in the blanks by using the clues the spreadsheet provides for you. This data is from *The World Almanac and Book of Facts*.

The Principal Languages of the World

	A	B	C	D
	Language	**Number of Native Speakers (millions)**	**Number of Non-native Speakers (millions)**	**Total Number of Speakers (millions)**
1				
2	Mandarin	844		975
3	English	326		478
4	Hindi		97	437
5	Spanish	339		392
6	Russian	169		284
7	Arabic	190	35	
8	Bengali	193	7	
9	Portuguese	172		184
10	Malay-Indonesian		107	159
11	Japanese	125		126
12	French		52	125
13	German	98		123
14	TOTAL			

2. Write a spreadsheet formula that will calculate the value for D3.

3. Write a spreadsheet formula that will calculate the value for B14.

4. What is the total number of people who speak these 12 languages?

Extending Concepts

Use only these words and mathematical symbols to solve the following problems.

- **Words:** Profit (P), Income (I), Expenses (E), Cost (C), Price (R), Number Sold (S)
- **Symbols:** $+, -, \times, \div, =$

5. If Profit = Income − Expenses, or P = I − E, use words and symbols to write another way of calculating Income.

6. If Profit = Income − Expenses, or P = I − E, use words and symbols to write another way of calculating Expenses.

Making Connections

Below is a spreadsheet about three qualities of animals: gestation, average longevity, and maximum speed. Use this information from *The World Almanac and Book of Facts 1996* to answer the questions below.

	A	B	C	D
	Animal	**Gestation (days)**	**Average Longevity (years)**	**Maximum Speed (mph)**
1				
2	Grizzly Bear	225	25	30
3	Cat (domestic)	63	12	30
4	White-tailed Deer	201	8	30
5	Giraffe	425	10	32
6	Lion	100	15	50
7	Zebra	365	15	40

7. Write a spreadsheet formula to calculate the average gestation period for the 6 animals.

8. If the maximum speed for a human is 27.89 mph, by how much does it differ from the average maximum speed of the 6 animals?

What relationships exist among two-dimensional figures?

GETTING IN

SHAPE

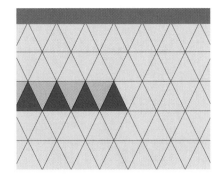

PHASE**ONE**
Triangles

In this phase, you will estimate angle measures and draw and measure different types of angles. You will also investigate various angle and side relationships in triangles, explore the different ways you can classify triangles, check for lines of symmetry, and explore similarity in triangles. Finally, you will enlarge a logo and determine what is involved in creating similar triangles.

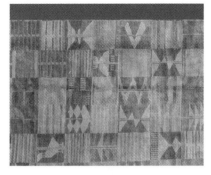

PHASE**TWO**
Polygons

You will develop your own classification system for sorting a set of Polygon Tiles™ and explore how different four-sided figures are related. Then you will investigate angle relationships in polygons and discover your own formulas. You will expand your knowledge of congruence to include polygons and explore how to use transformations to determine whether polygons are congruent. You conclude by investigating congruence using transformations on the coordinate plane.

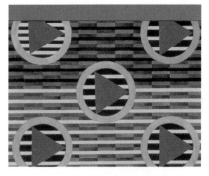

PHASE**THREE**
Circles

You will investigate the unique properties of circles and discover how the circumference and the diameter of a circle are related. The activities that follow will lead you to develop formulas for finding the area of a regular polygon and the area of a circle based on the familiar formula for the area of a rectangle. Finally, you will create a design using figures you have studied in this unit and write a report that explains the mathematical relationships behind your design.

PHASE ONE

In this phase, you will explore triangles by examining angles and sides. You will develop methods for describing and comparing triangles, including how to determine whether triangles are similar or not.

Logos often include geometric shapes such as triangles. Maintaining an exact design when replicating a logo is important. You will apply what you have learned about triangles to enlarge a logo while maintaining the shape and proportions of the original design.

Triangles

WHAT'S THE MATH?

Investigations in this section focus on:

GEOMETRY and MEASUREMENT

- Measuring and classifying angles
- Investigating the angles and sides of triangles
- Classifying triangles
- Finding lines of symmetry
- Identifying similarity in triangles

NUMBER

- Counting lines of symmetry

STATISTICS and PROBABILITY

- Recording angle measures and looking for patterns in your data
- Writing true statements about triangles

MathScape Online
mathscape2.com/self_check_quiz

1 The Angle on Angles

When you study geometry, angles are a good place to start. In this lesson, you begin your exploration of two-dimensional figures by creating an angle-maker and using it along with a protractor to investigate angle measurement and angle classification.

Estimate and Measure Angles

How closely can you estimate angle measures?

Follow the directions to create an angle-maker. For these activities, switch roles with your partner after three turns.

1. See how accurately you and your partner can make an angle.

   a. Name an angle measure between 0° and 360° and record the angle.

   b. Have your partner use the angle-maker to show the angle.

   c. Measure and record your partner's angle.

   d. Use a protractor and a straightedge to draw the angle.

2. Try it the other way. Use the angle-maker to show an angle.

   a. Have your partner estimate and record this measure.

   b. Check your partner's estimate with a protractor. Then draw the angle with a protractor.

How to Create an Angle-Maker

1. On each of three sheets of different-colored construction paper, use a compass to draw a large circle about 9 inches across. Then cut out the circles.

2. Use a straightedge to draw a line from the center of each circle to the edge. Then cut along the line to make a slit in each circle.

3. Fit two of the circles together along the slits as shown.

Investigate Three Angles in a Circle

Use an angle-maker with three colors to investigate how many ways three angles can be combined to fill a circle. Here is one possibility.

∠UAV	Acute
∠VAW	Acute
∠WAU	Reflex

Keep a written record of all the combinations you find. For each combination, record your results as follows.

- Draw a picture of how the angle-maker looks. Label the angles using letters of your choice.

- Record the names of the three angles that fill the circle.

- Classify each angle.

How many different ways can three angles be combined to fill a circle?

How to Name and Classify Angles

Angles are named by the point at their vertex. The angle shown is angle A, or ∠A. Sometimes it's helpful to use three letters to name angles. In this case, the vertex is the letter in the middle. We can call this angle ∠BAC or ∠CAB.

Angles are classified by their measures.

∠DEF is **acute.** It measures between 0° and 90°.

∠JKL is a **right** angle. It measures 90°.

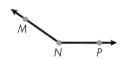

∠MNP is **obtuse.** It measures between 90° and 180°.

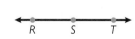

∠RST is a **straight** angle. It measures 180°.

∠XYZ is a **reflex** angle. It measures between 180° and 360°.

hot **words** | angle
vertex

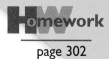

Homework

page 302

2

The Truth About Triangles

Triangles are geometric figures with three angles and three sides. Can any three angles be the angles of a triangle? Can any three sides be put together to form a triangle? Here's a way to investigate these questions.

Explore the Angles of a Triangle

What three angles of a triangle are possible?

Can the three angles of a triangle have any measures, or are there some restrictions? Use the triangular Polygon Tiles and the following ideas to guide your investigation.

1 Trace each triangular Polygon Tile on a sheet of paper.

2 Label the angles and measure them with a protractor. You may need to extend some of the sides so they are long enough to measure.

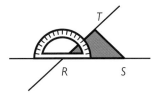

3 Record the angle measures, and then look for patterns in your data. Be sure to consider largest possible angles, smallest possible angles, the sum of the angle measures, and so on.

4 Write down any conclusions you can make about the angles of a triangle.

Explore the Sides of a Triangle

Can any three side lengths be put together to make a triangle? If not, how can you tell whether three given side lengths will form a triangle? You can investigate this using strips of paper as follows.

What can you say for sure about the sides of a triangle?

- Cut strips from a sheet of notebook paper. Each strip should be as wide as one line on the paper. Cut strips of these lengths: 3 cm, 4 cm, 5 cm, 6 cm, 7 cm, 8 cm, 9 cm, and 10 cm. Label each strip with its length.

- Choose three strips and place them together to see if they form a triangle.

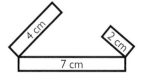 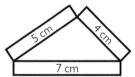

- As you experiment, keep a record of which strips you used and whether or not they formed a triangle.

- Write down any conclusions you can make about the sides of a triangle.

hot **words** | triangle

Hom**e**work
page 303

3 Can a Triangle Have Four Sides?

Like angles, triangles can be classified in different ways.
Knowing the different types of triangles makes it easier to talk about them and investigate what is and is not possible.

Explore the Possibilities

What combinations of side and angle classifications are possible for a triangle?

There are three ways to classify triangles by their sides and four ways to classify triangles by their angles.

1. Choose a pair of triangle classifications, such as scalene and obtuse, to investigate. Is it possible to have such a triangle?

2. Use a ruler, a protractor, and/or the triangular Polygon Tiles to find out whether such a triangle is possible. If it is possible, draw an example. If it is not possible, write an explanation.

3. Explore all of the combinations. (Be sure you have them all!) You may want to record your work in a chart.

Types of Triangles

Triangles can be classified by their sides.

Scalene	Isosceles	Equilateral
In a scalene triangle, no sides have the same length.	An isosceles triangle has at least two equal sides.	An equilateral triangle has three equal sides.

Triangles can also be classified by their angles.

Acute	Right	Obtuse	Equiangular
An acute triangle has three acute angles.	A right triangle contains a right angle.	An obtuse triangle contains an obtuse angle.	An equiangular triangle has three equal angles.

Relate Symmetry to Triangles

Is it possible to draw a triangle with the following number of lines of symmetry? If so, draw an example and tell what type of triangle it must be.

1 No lines of symmetry

2 Exactly one line of symmetry

3 Exactly two lines of symmetry

4 Exactly three lines of symmetry

5 Four or more lines of symmetry

What can you say about the lines of symmetry for different types of triangles?

Looking for Symmetry

A figure has *symmetry* (or is *symmetric*) if there is at least one line that divides it into two halves that are mirror images of each other.

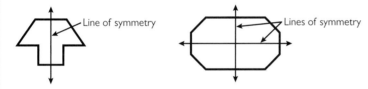

Write True Triangle Statements

Using the language and ideas of this lesson, write as many true statements about triangles as you can. Here are some examples of good statement openers to get you started.

- All isosceles triangles have . . .

- Every scalene triangle . . .

- No right triangle is also . . .

- Every triangle with three lines of symmetry . . .

hot **words** | symmetry
line of symmetry

page 304

Enlarging Triangles

In this lesson, you will learn some terms that are used to compare triangles. Two triangles are called **similar triangles** if one looks like an exact enlargement of the other. How do you decide whether two triangles are similar?

Create an Enlargement

What are some methods for enlarging triangles?

The Student Council at Fort Couch Middle School has been asked to help decorate the Triangle Café by making posters of its logo. Follow the instructions

to make an enlargement of the logo using grid paper that your teacher will provide. After you have made the enlargement, answer the questions below.

1 How does your enlarged logo compare to the original?

2 Are the angles the same? Are the lengths of the sides the same? Is it the same type of triangle?

3 The Student Council has requested that you put an enlarged copy of the logo on one wall of the café. Write down as many ways as you can think of to accomplish this task.

Making a Grid Enlargement

Step 1 Carefully examine the logo on the small grid paper and choose one box to use as a starting point.

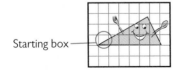

Starting box

Step 2 Find the corresponding box on the larger grid paper.

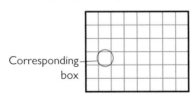

Corresponding box

Step 3 Copy the starting box so that it fits into the larger box as shown.

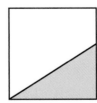

Step 4 Repeat Steps 1–3 until the entire figure has been enlarged onto the larger grid paper.

Explore Similarity

Using the handouts your teacher will provide, examine the enlarged triangles created by the class. Analyze how each triangle compares to the original triangle. After you have completed the chart, answer the following questions.

1 Which triangle or triangles are exact enlargements of the original logo?

2 What generalizations can you make about similar triangles?

What must be true for two triangles to be called *similar* triangles?

Definitions

Corresponding parts are parts on the enlarged figure that match the original figure. For example, angle *A* and angle *D* are corresponding parts. Line segments *AB* and *DE* are corresponding parts. Can you name other corresponding parts?

Congruent angles are angles that have exactly the same measure.

Congruent triangles are exact copies of the same triangle.

Letter to Students

Choose a student whose triangle is not similar to the original logo and write a letter to him or her explaining why the triangle does not meet the Student Council's request. In your letter, be sure to discuss the following:

- Explain the meaning of similarity.

- Explain how the triangle does not meet the definition of a similar triangle.

- Be specific about which parts of the triangle are drawn incorrectly.

- Use the appropriate terminology to describe the triangles. For example, specify scalene, isosceles, or equilateral. Also specify acute, right, obtuse, or equiangular.

hot **words** | similar figures

Homework
page 305

PHASE TWO

In this phase, you will make some fascinating discoveries about polygons. You will find out how different types of four-sided figures are related, explore angle relationships, and look into the relationship between symmetry and regular polygons. Finally, you will explore congruence.

Professionals who prepare blueprints rely on their knowledge of geometry to display in detail the technical plans architects and engineers draw up for their clients.

Polygons

WHAT'S THE MATH?

Investigations in this section focus on:

GEOMETRY and MEASUREMENT

- Exploring different types of quadrilaterals

- Creating tessellations

- Finding lines of symmetry in polygons

- Analyzing the relationship between symmetry and regular polygons

- Exploring congruence using flips, turns, and slides

- Investigating flips (reflections), turns (rotations), and slides (translations) on the coordinate plane

- Writing about transformations

STATISTICS and PROBABILITY

- Developing a classification system for polygons

- Keeping an organized list of results

ALGEBRA and FUNCTIONS

- Writing a formula for angle relationships in polygons

MathScape Online

mathscape2.com/self_check_quiz

5 Polygon Power!

A polygon is a closed figure, in a plane, with three or more sides. You already know about triangles—the simplest polygons. It makes sense to ask some of the same questions about other types of polygons. Classifying figures is always a good place to start; then you can begin investigating what is and is not possible.

Sort Polygons

What are some different ways that polygons can be classified?

Sorting polygons helps you develop your own ways of classifying them. Begin with a complete set of Polygon Tiles. Work with classmates to sort them in three different ways that make sense to you.

Be sure to take notes on how you sort the Polygon Tiles, since you may be asked to describe your methods to the class.

Number of Angles

3	4	5	6	7

Types of Quadrilaterals

Here are some types of quadrilaterals you should know.

Parallelogram

A parallelogram is a quadrilateral with two pairs of parallel sides.

Rectangle

A rectangle is a quadrilateral with four right angles.

Square

A square is a quadrilateral with four right angles and all sides of equal length.

Rhombus

A rhombus is a parallelogram with all sides of equal length.

Trapezoid

A trapezoid is a quadrilateral with exactly one pair of parallel sides.

Isosceles Trapezoid

An isosceles trapezoid is a trapezoid with nonparallel sides of equal length.

Investigate Quadrilateral Relationships

As you explore the following with classmates, keep track of your results using words and/or drawings.

How are the different types of quadrilaterals related?

1 Find all the quadrilateral Polygon Tiles that fit each definition. Which definitions have the most Polygon Tiles? the fewest?

2 Now try it the other way around. For each quadrilateral Polygon Tile, find all the definitions that apply to it. Which Polygon Tiles have the most definitions? the fewest?

3 Write as many statements as you can about the relationships among quadrilaterals. For example, "Every square is also a. . . ."

Summarize Quadrilateral Relationships

Find a way, such as a "family tree," Venn diagram, or some other method, to summarize the quadrilateral relationships you examined. Explain how your summary works.

hot **words** | polygon
quadrilateral

HW **omework**

page 306

Standing in the Corner

The sum of the angles of any triangle is 180°. Can you say something similar about the sum of the angles of other polygons? First you will investigate this for quadrilaterals; then you will look for patterns that work for any polygon.

Explore Angles of Quadrilaterals

What can you say about the sum of the angles of a quadrilateral?

Use the four-sided Polygon Tiles to find out about the sum of the angles for any quadrilateral.

1 Choose a four-sided Polygon Tile and carefully trace it on a sheet of blank paper.

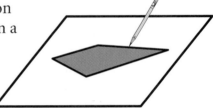

2 Extend the sides of the quadrilateral so you can measure the angles with a protractor.

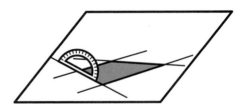

3 Measure the four angles and calculate their sum. Record your findings.

4 Repeat the process with other four-sided Polygon Tiles until you are ready to make a generalization.

Explore the Angles of Any Polygon

You know that three-sided polygons (triangles) have angles that add up to 180°, and you just investigated this for polygons with four sides (quadrilaterals). What happens with polygons that have five or more sides?

What can you say about the sum of the angles of polygons in general?

Each member of your group should choose a number from those shown below. Everyone should have his or her own number. You will be responsible for investigating the sum of the angles of polygons with the number of sides you choose.

5	6	7	8	9	10

1. Choose a Polygon Tile that has the number of sides you are investigating and carefully trace it on a sheet of blank paper. You can also draw your own polygon.

2. Extend the sides of the polygon so you can measure the angles with a protractor.

3. Measure the angles and calculate their sum. Record your findings.

4. Repeat the process with other polygons that have your chosen number of sides until you are ready to make a generalization about the sum of their angles.

Look for a Pattern

Write a description of any patterns you see in the record you kept. Use words, formulas, and/or equations.

Once you know the number of sides, you should be able to use your results to find the sum of the angles of any polygon.

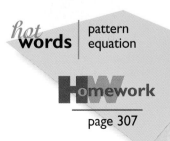

hot **words** | pattern
equation

Homework
page 307

7 Moving Polygons Around

You've already learned about congruent triangles. What do you think it means if two *polygons* are congruent? How can you show that two polygons are congruent? In this lesson you will explore ways to determine whether polygons are congruent.

Explore Congruence

How can you decide whether polygons are congruent?

Look at the cards your teacher has given you and compare them to each of the figures below. Decide whether each figure on the card is congruent to any of the figures below. Be prepared to explain your reasoning.

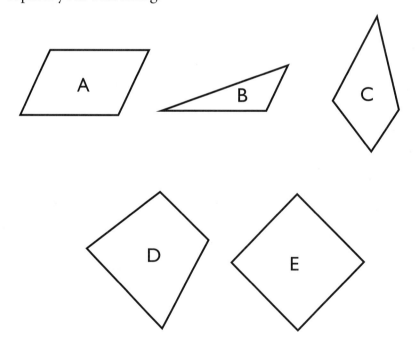

Investigate Polygons on the Coordinate Plane

Use what you know about flips, turns, and slides to answer each question. Refer to the handout provided by your teacher.

1 Write the coordinates for each vertex of figure 1 and figure 2.

2 Describe the move that occurred from figure 1 to figure 2 as a *slide*, *flip*, or *turn*.

3 Predict the coordinates for each vertex in figure 2 if it is shifted up 2 units. Check your prediction by graphing the figure.

4 Write a general rule for how the coordinates (ordered pairs) of the vertices change for vertical slides.

5 Write the coordinates for each vertex of figure 3.

6 Describe the move that occurred from figure 2 to figure 3.

7 Predict the coordinates for each vertex in figure 3 if it is shifted 4 units to the right. Check your prediction by graphing the figure.

8 Write a general rule for how coordinates (ordered pairs) change for horizontal slides.

9 What move happened from figure 4 to figure 5?

10 What move happened from figure 6 to figure 7?

How does the coordinate plane help you determine whether two polygons are congruent?

Writing Transformations

Write directions telling your partner how to transform figure 4. You must include at least one vertical and one horizontal slide. Graph the transformation on your own before you give the directions to your partner. Exchange directions with your partner and follow each other's directions. Check your partner's work using your graph.

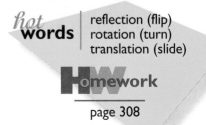

hot **words** | reflection (flip)
rotation (turn)
translation (slide)

Homework

page 308

8 Symmetric Situations

You have probably noticed that some polygons look nice and even, while others look dented or lopsided. How a polygon looks often has to do with symmetry. In this lesson, you will explore this connection and see how symmetry is related to regular polygons.

Explore Lines of Symmetry

How are lines of symmetry related to the shape of a polygon?

Use Polygon Tiles to explore the lines of symmetry of polygons. Follow these steps.

1 Choose a Polygon Tile and trace it.

2 Use a straightedge to draw all of the polygon's lines of symmetry.

3 Repeat this with other polygons, and keep track of your results. Try a wide variety of polygons, and look for patterns in your results.

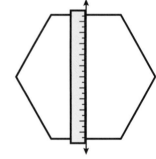

What can you say about the lines of symmetry of regular polygons?

Write About Polygons

How are two polygons alike, and how are they different?

Your teacher will provide you with two cards, each of which has a pair of polygons pictured on it. For each card, prepare a report using the directions that follow. Whenever possible, include illustrations.

- Classify each polygon in as many ways as possible using all the terminology you know about polygons. For example, what is the name of the polygon? Is it regular, isosceles, scalene, equilateral, concave, or convex?

- Find the sum of the angles of each polygon. Include anything else you can say about the angles.

- Tell how many lines of symmetry each polygon has.

- Are the polygons similar to each other? Describe why or why not, using corresponding parts of the polygons.

- Is one polygon related to the other by a flip, turn, or slide? Explain in detail.

- Are the polygons congruent? Describe how you know.

Be sure to label each of your two reports with the option letter and card numbers, so it is clear which polygons you are comparing. In addition, staple your cards to each report when you turn them in.

hot **words** | symmetry
line of symmetry

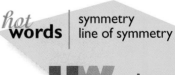

page 309

PHASE THREE

In this phase, you will come "full circle" in your exploration of two-dimensional figures. After investigating a relationship that makes circles special, you will find out how circles and polygons are related. Then you will create a geometric design based on the mathematical ideas in this unit.

Art museums throughout the world have abstract art collections. Many of these works of art use geometric ideas presented in vivid colors.

Circles

WHAT'S THE MATH?

Investigations in this section focus on:

GEOMETRY and MEASUREMENT

- Measuring the diameter and circumference of circular objects
- Forming rectangles from other polygons
- Investigating the relationship between the area of a rectangle and the area of a circle
- Creating a geometric design

ALGEBRA and FUNCTIONS

- Writing an equation for the relationship between diameter and circumference
- Developing and using formulas to find the area of a polygon and the area of a circle

STATISTICS and PROBABILITY

- Collecting, displaying, and analyzing measurement data

MathScape Online
mathscape2.com/self_check_quiz

Going Around in Circles

EXAMINING
CIRCLES,
CIRCUMFERENCE,
AND PI

Circles are the next stop on your exploration of two-dimensional shapes. You will soon see how circles are related to the polygons you have been working with. For now, you will investigate a fascinating relationship between two of the measurements of any circle.

Measure Circles

How can you measure the diameter and circumference of a circle?

- Your goal is to measure the diameter and circumference of some circular objects as accurately as possible.

- You may use string, a measuring tape, a meterstick, or other tools to help you. All of your measurements should be in centimeters.

- Record your results in a table like the one shown here.

Object	Diameter	Circumference
Jar lid	10.4 cm	32.7 cm

- Be ready to describe your measurement methods to the class.

Parts of a Circle

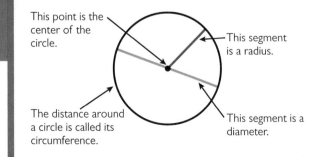

This point is the center of the circle.

This segment is a radius.

The distance around a circle is called its circumference.

This segment is a diameter.

Sometimes *radius* refers to the length of a radius of a circle. The radius of this circle is 1.5 cm. Similarly, *diameter* may refer to the length of a diameter of a circle. The diameter of this circle is 3 cm.

Analyze the Data for Circle Relationships

Use your measurement data to analyze the relationship between the diameter and the circumference of a circle.

- Describe in words how the diameters and circumferences you measured are related.

- Make a scatterplot on a coordinate grid for your pairs of data points. Describe any relationships you see.

- Describe the relationship between diameter and circumference with an equation. Be as accurate as you can.

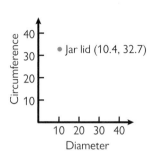

What is the relationship between diameter and circumference?

Apply Circle Relationships

The radius of Earth is approximately 4,000 miles. Write a brief paragraph outlining what this fact tells you about Earth's measurements, including diameter and circumference. Also provide an explanation of how you found your results.

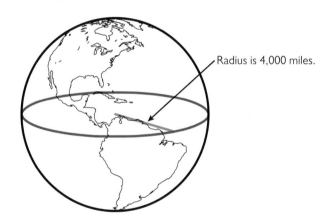

Radius is 4,000 miles.

hot **words** | circumference pi

Homework

page 310

10 Rectangles from Polygons

Before tackling the area of circles, it is helpful to investigate the area of regular polygons. As you will see, regular polygons can be divided up into triangles. Thinking of regular polygons in this way allows you to discover a formula for their area.

Turn Polygons into Rectangles

Can any regular polygon be rearranged to form a rectangle?

One of the special things about regular polygons is that you can divide them into wedges of the same size and shape. To do this, you start at the center and draw a line to each vertex of the regular polygon. If you divide a regular polygon into wedges like this, you can arrange the wedges to form a rectangle. Your goal is to find a way to rearrange any regular polygon so that it forms a rectangle.

1 Choose a Polygon Tile that is a regular hexagon or regular octagon. Trace it on a sheet of paper, and then cut out the polygon.

2 Draw lines from the center of the polygon to each vertex. Cut out the wedges.

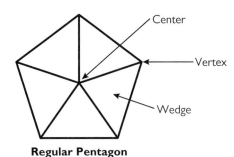

Regular Pentagon

3 Find a way to rearrange the wedges to form a rectangle. You may need to cut one or more of the wedges into smaller pieces.

4 Make a drawing that shows how you used the pieces of the polygon to form a rectangle.

Develop a Formula for the Area of a Regular Polygon

Look back at the rectangle you formed from the pieces of the regular polygon. Then discuss these questions with classmates.

- How is the area of the rectangle related to the area of the regular polygon?

- How can you find the area of the rectangle?

- How does the length of the rectangle relate to the regular polygon?

- How does the width of the rectangle relate to the regular polygon?

- What formula would help you find the area of the regular polygon?

Can you come up with a formula for the area of a regular polygon?

Try Out the Formula

Now it's time to put your formula to the test. Keep a record of your work as you follow these steps.

1 Choose a regular Polygon Tile and measure one side of it.

2 Find the perimeter of the Polygon Tile.

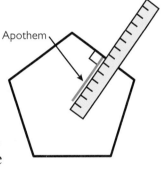
Apothem

3 Locate the center of the tile as accurately as you can. Then measure the apothem of the tile. The **apothem** is the perpendicular distance from the center to a side.

4 Use your measurements and your formula for the area of a regular polygon to find the area of your Polygon Tile. Be sure to include units in your answer.

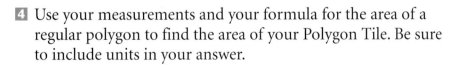

hot **words** | rectangle
                    area

Homework
page 311

11 Around the Area

Now that you have found the area of a regular polygon, you will see that finding the area of a circle is not so different. After estimating the area of a circle, you will develop a formula for the area of a circle using a method similar to the one you used for a regular polygon. Then you will apply the formula to see how good your original estimate was.

Estimate the Area of a Circle

How can you use centimeter grid paper to estimate the area of a circle?

Work with a partner to estimate the area of a circle. Use the following steps.

1 Choose a radius for your circle. Use either 5 cm, 6 cm, 7 cm, 8 cm, or 9 cm.

2 Use a compass to draw a circle with this radius on centimeter grid paper.

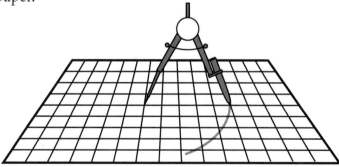

3 Find a way to estimate the area of your circle. Make as accurate an estimate as you can. Keep a written record of your results and be ready to share your method with the class.

Develop a Formula for the Area of a Circle

Follow the steps for turning a circle into a rectangle. Sketch your results. Then get together with classmates to discuss these questions.

- How is the area of this "rectangle" related to the area of the circle?

- If your "rectangle" were a true rectangle, how could you find its area?

- How does the length of the "rectangle" relate to the circle?

- How does the width of the "rectangle" relate to the circle?

- How can you write a formula for the area of the circle?

Can you find a formula for the area of a circle?

How to Turn a Circle into a Rectangle

1. Use a compass to draw a large circle on a sheet of paper. Cut out the circle.

2. Fold the circle in half three times. When you unfold the circle, it will be divided into equal parts or sectors.

3. Cut along the fold lines to separate the sectors.

4. Arrange the sectors so that they form a figure that is as close to a rectangle as possible.

Use the Formula to Evaluate Estimates

Write a brief summary of your findings in this lesson. Include the following:

- the formula for the area of a circle

- the area of the circle you drew at the beginning of the lesson, calculated using the formula

- a comparison of your original estimate and the calculated area

hot **words** | radius

H▪Womework

page 312

12 Drawing on What You Know

From the repeating patterns that decorate our clothing to the tessellations created with floor tiles, geometric designs are all around us. Behind every geometric design are some mathematical properties that you have explored in this unit. This lesson gives you an opportunity to illustrate these mathematical ideas by creating a geometric design of your own.

Create a Geometric Design

How can you make a geometric design that shows what you know about triangles, polygons, and circles?

It's your turn to create an original geometric design. You may use a ruler, protractor, compass, Polygon Tiles, or any other tools. Follow these guidelines:

- Your design should illustrate at least five mathematical concepts from this unit. Include more mathematical concepts if you can.

- Your design must include triangles, polygons with more than three sides, and at least one large circle.

- You may use color in your design.

Geometric Designs

These photographs are close-ups of a honeycomb and a piece of woven cloth from Africa.

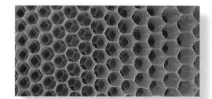

Honeycomb

Cloth pattern

Write a Report on Your Design

Prepare a report to accompany your geometric design. Here are some guidelines on what to include:

- An overall description of your design and the tools you used to create it

- A description of all of the mathematical ideas in your design with additional drawings, if necessary, to explain the mathematics

Be sure to use the correct vocabulary as you describe the geometric ideas in your design.

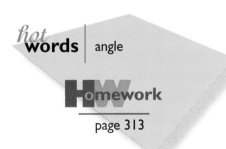

hot**words** | angle

Homework
page 313

The Angle on Angles

Applying Skills

Estimate the measure of each angle.

1. **2.**

3. **4.**

5. Use your protractor to measure the angles in items **1–4**.

6. Make a table comparing your estimates in items **1–4** with your measurements in item **5**.

Name each angle in three different ways. Classify the angle as acute, right, obtuse, straight, or reflex.

7. **8.**

9. 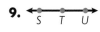 **10.**

Use a protractor and straightedge to draw an angle with each measure. Label each angle with three letters.

11. $\angle KLM$, 38°   **12.** $\angle CDE$, 70°

13. $\angle XYZ$, 150°   **14.** $\angle QRS$, 95°

Extending Concepts

There are eight ways to combine three angles to fill a circle. One way is to use two acute angles and a reflex angle as shown.

15. How many ways can three angles be combined to form a straight angle? For each possibility draw a picture and classify the three angles.

16. Suppose that four acute angles are combined to form a straight angle as shown. What is the total number of acute angles formed? obtuse angles? reflex angles? List the names of the angles of each type.

Making Connections

The pie chart shows the breakdown by age of the population of the United States in 1990. The size of each sector is proportional to the percentage in that category.

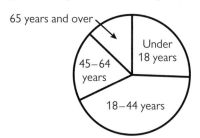

17. For each sector, measure the angle using your protractor. Classify the angle as acute, obtuse, reflex, or right.

18. Calculate the percentage of the population in each age group. Explain how you calculated the percentages.

The Truth About Triangles

Applying Skills

Two of the angle measures of a triangle are given. Find the measure of the third angle.

1. 25°, 80°
2. 65°, 72°

3. 105°, 8°
4. 12°, 14°

Tell whether each set of side lengths could be used to form a triangle.

5. 7, 9, 10
6. 4, 3, 10

7. 2.4, 5.7, 8.0
8. 10.8, 13.9, 3.1

Two side lengths of a triangle are given. What can you say about the length of the third side?

9. 8 in., 15 in.
10. 21 cm, 47 cm

11. 2.1 m, 2.8 m
12. 10 ft, 14 ft

13. Sketch a triangle with angles 59°, 60°, and 61°. Label the angle measures. Put an X on the longest side.

Extending Concepts

14. A triangle has two angles with the same measure.

a. What can you say about the lengths of the sides opposite the two equal angle measures?

b. Could the two angles with the same measure be right angles? obtuse angles? Explain your thinking.

15. The post office is 100 yards from the bank and 50 yards from the store. The bank, store, and post office form a triangle.

a. What can you say about the distance between the store and the bank?

b. The library is equally far from the bank and the post office. If the library, bank, and post office form a triangle, what can you say about the distance between the library and the bank?

c. Which is shorter, walking directly from the post office to the bank or walking first from the post office to the store and from there to the bank? How does this relate to what you have learned in this lesson about the sides of a triangle? Draw a sketch and write an inequality.

Writing

16. Answer the letter to Dr. Math.

Dear Dr. Math:

My friend Lena said that a 179° angle is the largest possible angle in a triangle. But I know that's not right because you can have a $179\frac{1}{2}$° angle. Then my friend Alex said well how about a $179\frac{3}{4}$° angle? or a $179\frac{7}{8}$° angle? Now I'm confused. Can you tell us what the maximum possible angle in a triangle is?

Angus in Angleside

Can a Triangle Have Four Sides?

Applying Skills

Classify each triangle by its sides and by its angles.

1.

2.

3.

4.

Tell whether each statement is true or false.

5. Every obtuse triangle is a scalene triangle.

6. Every isosceles triangle has at least one line of symmetry.

7. No right triangle has three lines of symmetry.

Tell how many lines of symmetry each figure has.

8.

9.

10.

11.

Sketch an example of each type of triangle.

12. right, isosceles   13. scalene, obtuse

14. acute, one line of symmetry

Extending Concepts

15. A triangle can be classified by its angles as acute, right, obtuse, or equiangular.

    a. For each, tell which is possible: no lines of symmetry, one line of symmetry, three lines of symmetry.

    b. Draw an example of each type and show the lines of symmetry.

The *converse* of a statement "If A, then B" is "If B, then A." Tell whether each statement below is true or false. Then write the converse statement and tell whether the converse is true or false.

16. "If a triangle is an equilateral triangle, then it is an acute triangle."

17. "If a triangle is an isosceles triangle, then it is an equilateral triangle."

Making Connections

18. The Ashanti people of western Africa used brass weights to measure gold dust, which was their currency. Each weight represented a local proverb. The picture shows an example of one such weight. Classify the inner triangle by its sides and by its angles. How many lines of symmetry does it have? How many lines of symmetry does the weight as a whole have?

Enlarging Triangles

Applying Skills

Determine whether the following pairs of triangles are congruent. Write *yes* or *no*.

1.

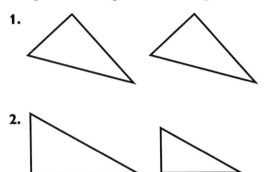

2.

For items 3–6, tell whether each statement is *true* or *false*.

3. Congruent angles have exactly the same measure.

4. Angles are congruent if they are within 5 degrees of one another in measure.

5. A right angle is congruent to a straight angle.

6. An angle that has a measure of 90 degrees is congruent to a right angle.

7. Name each pair of corresponding sides and corresponding angles for the two similar triangles.

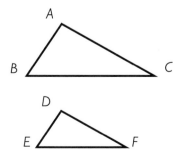

For items 8 and 9, decide whether the following pairs of triangles are similar or not. Describe why or why not.

8.

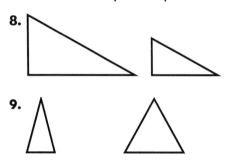

9.

Extending Concepts

10. Use your protractor for this problem. Are any of these overlapping triangles similar? Explain your answer.

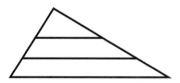

Writing

11. Suppose your friend drew the two triangles below and told you they are similar. Write a note to your friend explaining whether or not you think they are similar. Give at least two reasons to support your answer.

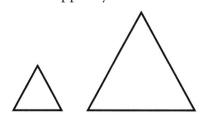

Polygon Power!

Applying Skills

For each polygon, state the number of sides, name the polygon, and tell whether it is convex or concave.

1.    **2.**

Classify each quadrilateral. Give all the names that apply to it.

3.   **4.**

5. Sketch an example of a concave pentagon and a convex heptagon.

Tell whether each statement is true or false.

6. Every rectangle is also a parallelogram.

7. Every parallelogram is also a rhombus.

8. Some trapezoids have two pairs of parallel sides.

For each definition, sketch an example and name the figure.

9. A parallelogram with four right angles

10. A quadrilateral with exactly one pair of parallel sides

Extending Concepts

Tell whether each statement is true or false. If it is false, sketch a counterexample.

11. If a quadrilateral is not a parallelogram, then it is a trapezoid.

12. If a trapezoid has two sides of equal length, it is an isosceles trapezoid.

The *inverse* of a conditional statement is obtained by negating both parts. For example, the inverse of the statement "If a polygon is a pentagon, then it has five sides" is "If a polygon is *not* a pentagon, then it *does not have* five sides."

13. Tell whether the statement "If a quadrilateral is a square, then it is also a rhombus" is true or false. Then write the inverse statement and tell whether it is true or false.

Making Connections

Native American tribes traditionally used beads to decorate clothing, blankets, and bags. The picture shows an example of beadwork created by an Eastern Woodland tribe to decorate a bandolier shoulder bag.

14. Identify as many different polygons as you can. Name each polygon and tell whether it is convex or concave. If you identify any quadrilaterals, tell whether they are one of the special types of quadrilateral.

Standing in the Corner

Applying Skills

Find the sum of the angles of a polygon with:

1. 6 sides   **2.** 8 sides   **3.** 28 sides

Find the sum of the angles and the measure of each angle of a regular polygon with:

4. 7 sides   **5.** 9 sides   **6.** 55 sides

Find the measure of each angle of:

7. a regular octagon   **8.** a regular pentagon

Find the sum of the angles of each polygon.

9.    **10.**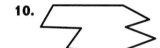

Extending Concepts

11. The sum of the angles of a polygon with n sides is $S = (n - 2)180°$. How many sides does a polygon have if the sum of its angles is 1,440°?

12. Sketch a convex pentagon. From one vertex draw a diagonal to every other nonadjacent vertex of the pentagon. How many triangles are formed?

13. Repeat the process used in item **12** for a hexagon, a heptagon, and an octagon. How many triangles are formed in each case?

Let n be the number of sides of a polygon and T the number of triangles that are formed when diagonals are drawn from one vertex.

14. Describe in words the relationship between n and T. Write an equation that relates n and T.

15. The sum of the angles of a polygon with n sides is $S = (n - 2)180°$. Use your result from item **14** to explain why this formula makes sense.

Writing

16. Answer the letter to Dr. Math.

> **Dear Dr. Math:**
>
> I used the formula $S = (n - 2)180°$ to find the sum of the angles for this polygon. It has five sides, so I multiplied $180°$ by 3 to get $540°$. To check my answer, I measured the angles and got $90°, 90°, 90°, 45°,$ and $45°$. When I added these angle measures, I got $360°$, not $540°$. Help! What did I do wrong?
>
> **Polly Gone**

Moving Polygons Around

Applying Skills

Tell whether each pair of figures is congruent.

1.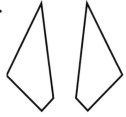

2.

3.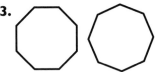

4.

Describe the transformation that has taken place for each figure on the coordinate plane.

5. figure 1 to figure 2

6. figure 3 to figure 4

7. figure 5 to figure 6

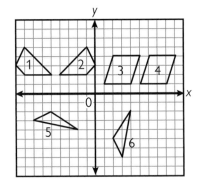

Extending Concepts

For items 8 and 9, refer to figure 4 at the bottom of the first column.

8. Predict the coordinates for each vertex of figure 4 if it were to slide 3 units down.

9. On a piece of graph paper, draw a figure that is congruent to figure 4. Place one vertex at point (3, 3).

Making Connections

Inductive reasoning involves reaching a conclusion by observing what has happened in the past and looking for a pattern. *Deductive reasoning* involves reaching a conclusion by using logic. Each of the following arguments supports the conclusion that the position or orientation of two figures does *not* affect their congruence.

Jose says, "I took 5 polygons on the coordinate plane and transformed them using flips, turns, and slides. None of those changes affected the size or shape of the figure. I measured each one and they were still all the same size and shape. So transformations do not affect whether two figures are congruent."

Margo says, "Since all you are doing when you transform a figure is changing its position — you do not change its size or shape. Therefore, it makes sense that transformations do not affect the congruence of figures."

10. Which type of reasoning did each person use? Which argument seems more convincing to you? Why?

Symmetric Situations

Applying Skills

Tell how many lines of symmetry each polygon has.

1. ▢

2. (trapezoid)

3. (pentagon)

4. (arrow/chevron)

5. (house/pentagon)

6. (hexagon)

7. (bowtie/hourglass)

8. (octagon)

9. a regular pentagon

10. a regular polygon with 27 sides

Extending Concepts

11. Make a sketch of each complete polygon by drawing the mirror image of the first half along the line of symmetry.

a.

b.

Then write a report in which you:

- classify each polygon in as many ways as possible and find the sum of the angles

- tell how many lines of symmetry each polygon has

12. Is it possible to draw a pentagon with no lines of symmetry? exactly 1 line of symmetry? exactly 2, 3, 4, or 5 lines of symmetry? Find all the possibilities. Sketch an example of each and show the lines of symmetry.

13. Is it possible to draw a quadrilateral with exactly 2 lines of symmetry? a pentagon with exactly 2 lines of symmetry? a hexagon? a heptagon? In each case, if it is possible, sketch an example. In general, how can you tell whether it is possible to draw a polygon with *n* sides and exactly two lines of symmetry?

Making Connections

It is believed that no two snowflakes are exactly alike. Snow crystals usually have hexagonal patterns, often with very intricate shapes. Each crystal takes one of seven forms depending on the temperature. The pictures below show two of these forms, *plates* and *stellars*.

14. How many lines of symmetry does each crystal have?

Going Around in Circles

Applying Skills

For all problems, use 3.14 for the value of pi. Round answers to one decimal place if necessary.

1. Sketch a circle. Draw and label a diameter, a radius, and the center.

2. If the diameter of a circle is 44 in., what is its radius?

3. If the radius of a circle is 1.9 cm, what is its diameter?

Find the circumference of each circle below.

4. radius = 4

5. diameter = 9

6. radius = 12.8 cm

7. diameter = 18.7 ft

8.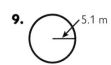

9.

Find the diameter and radius of a circle with the given circumference.

10. 3 m

11. 40.7 in.

12. 10.2 cm

13. 0.5 ft

Extending Concepts

14. a. Find the circumference of a circle with a diameter of 8 in. Use 3.14 for π. Is your answer exact or an approximation?

b. Most calculators give the value of π as 3.141592654. If you used this value of π in part **a**, would your answer be exact? Why or why not?

c. The circumference of a circle with a diameter of 8 in. could be expressed as 8π in., leaving the symbol π in the answer. Why do you think this might sometimes be an advantage?

15. a. Suppose that the radius of a large wheel is three times the radius of a small wheel. How is the circumference of the large wheel related to the circumference of the small wheel? How do you know?

b. If the radius of the small wheel is 10 in., what is its circumference? How far would it travel along the ground in one complete rotation?

Making Connections

16. a. The diameter of the moon is approximately 2,160 miles. What is its circumference? Give your answer to the nearest hundred miles.

b. The average distance of the moon from Earth is about 240,000 miles. Assuming that the moon travels in a circular orbit around Earth, about how far does the moon travel in one complete orbit? Give your answer to the nearest ten thousand miles. Explain how you solved this problem.

c. What are some reasons why your answer in part **b** is not exact?

Rectangles from Polygons

Applying Skills

Find the perimeter of each regular polygon.

1. octagon, side length = 5 cm

2. pentagon, side length = 10.5 in.

Find the area of each regular polygon. Round your answers to the nearest tenth.

3.

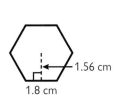

1.56 cm

1.8 cm

4.

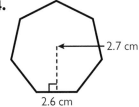

2.7 cm

2.6 cm

5. regular heptagon, side length = 4.0 ft, apothem = 4.15 ft

6. regular hexagon, side length = 5.4 cm, apothem = 4.68 cm

7. regular octagon, perimeter = 32 in., apothem = 4.83 in.

Extending Concepts

8. Use the regular octagon to answer the questions and draw sketches.

9.66

8

a. Show how the octagon can be divided into same-size triangles.

b. Show how the triangles can be rearranged to make a rectangle.

c. How does the length of the rectangle relate to the octagon? What is its length? How does the width of the rectangle relate to the octagon? What is its width?

d. How does the area of the rectangle relate to the area of the octagon? Why? What is the area of the rectangle? What is the area of the octagon?

9. a. Use the formula $A = l \cdot w$ to find the area of a square with side length 4.

b. What are the apothem and perimeter of a square with side length 4? How did you find them? Now use the formula $A = \frac{1}{2}ap$ to find the area of a square with side length 4. Is this the same as your answer in part **a**?

Writing

10. Answer the letter to Dr. Math.

> Dear Dr. Math:
>
> I wanted to find the area of this hexagon. I drew the apothem as I've shown. Then I measured the apothem and perimeter. I used the formula $A = \frac{1}{2}ap$ and got 3.2 cm^2 for the area, but that seems way too small to me.
> I think my teacher got lucky when she used that formula because it only works once in a while.
> Do you agree?
>
> K. L. Ooless

Around the Area

Applying Skills

Estimate the area of each circle. Then calculate the area. Use 3.14 for the value of π. Round your answers to one decimal place.

1. **2.** **3.**

4. Tell how the actual areas compare with your estimates.

Find the area of each circle. Use 3.14 for the value of π. Round your answers to one decimal place.

5. radius = 7

6. diameter = 13.5 in.

7. radius = 5 in.

8. radius = 4.1 cm

9.  2.6 cm

10.  1.5 m

Extending Concepts

11. a. A circle with a radius of 10 in. is divided into eight equal sectors. Make a sketch showing how you could rearrange the sectors to make a "rectangle."

b. How does the length of the "rectangle" relate to the circle? What is its length? How does the width of the "rectangle" relate to the circle? What is its width?

c. How does the area of the "rectangle" relate to the area of the circle? Why? If you use $A = l \cdot w$, what do you get for the area of the "rectangle"? What is the area of the circle?

d. If you wanted the figure to look more like a rectangle, what could you do differently when you divide the circle into sectors? Make a sketch to illustrate your answer. Do you think that it makes sense to use the formula $A = l \cdot w$ for the "rectangle" you made even though it is not an exact rectangle? Why or why not?

12. What is the area of the outer circle? of the inner circle? What is the area of the ring? How did you figure it out?

Making Connections

13. A *yurt* is a circular, domed, portable tent used by the nomadic Mongols of Siberia. What would the floor area be in a yurt with a radius of 10 feet?

Drawing on What You Know

Applying Skills

For each polygon used to create the tessellation, give the following information.

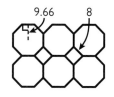
9.66   8

1. Name the polygon. For the quadrilateral, give every name that applies.

2. Find the sum of the angles and the measure of each angle.

3. Tell how many lines of symmetry the polygon has.

4. Find the perimeter and the area of the polygon to the nearest whole number.

Refer to the design to answer items 5–9.

5. Classify each angle of the triangle.

6. Classify the triangle by its sides and by its angles.

7. How many lines of symmetry does the triangle have?

8. What is the sum of the angle measures of the triangle?

9. Find the diameter, area, and circumference of the circle.

Extending Concepts

10. Create a geometric design of your own. Use a circle, a triangle, and a regular polygon that is not a triangle. Follow the guidelines to write a report on your design.

   • Describe your design and how you created it. Is it a tessellation?

   • Classify the triangle in as many ways as you can. Classify its angles and tell how many lines of symmetry it has.

   • Measure the radius of the circle. Calculate its diameter, area, and circumference.

   • Name the regular polygon, and find the sum of its angles. Measure its apothem and side length, and calculate its perimeter and area.

Writing

11. Answer the letter to Dr. Math.

Dear Dr. Math:

We've been learning all these formulas for areas of circles and polygons and for the sum of the angles of a polygon and lots of other things. But who cares about these formulas anyway? Why would I ever need to know the area of a circle? Does anyone ever use this stuff? If so, can you tell me who?

Fed Up in Feltham

GLOSSARY/GLOSARIO

Cómo usar el glosario en español:

1. Busca el término en inglés que desees encontrar.

2. El término en español, junto con la definición, se encuentra debajo del término en inglés.

A

absolute value a number's distance from zero on the number line

valor absoluto la distancia de un número desde cero en la recta numérica

Example/**Ejemplo:**

−2 is 2 units from 0.
−2 está a dos unidades de 0.

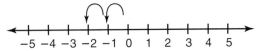

The *absolute value* of −2 is 2 or |−2| = 2.
El *valor absoluto* de −2 es 2 o |−2| = 2.

angle two rays that meet at a common endpoint

ángulo dos semirrectas que se encuentran en un punto final común

Example/**Ejemplo:**

$\angle ABC$ is formed by $\vec{BA}$ and $\vec{BC}$.
$\angle ABC$ está formado por $\vec{BA}$ y $\vec{BC}$.

area the size of a surface, expressed in square units

área el tamaño de una superficie expresada en unidades cuadradas

Example/**Ejemplo:**

2 ft
2 pie

area = 8 ft^2
área = 8 pie^2

4 ft
4 pie

B

base [1] the side or face on which a three-dimensional shape stands; [2] the number of characters a number system contains

base [1] el lado o cara donde descansa una figura tridimensional; [2] el número de caracteres que contiene un sistema numérico

benchmark a point of reference from which measurements can be made

punto de referencia un punto de referencia desde donde se pueden tomar medidas

budget a spending plan based on an estimate of income and expenses

presupuesto plan de gastos basado en una estimación de ingresos y expensas

C

cells small rectangles in a spreadsheet that hold information. Each rectangle can store a label, number, or formula

casillas rectángulos pequeños en una hoja de cálculo que contienen información. Cada rectángulo puede contener un rótulo, número o fórmula

chance the probability or likelihood of an occurrence, often expressed as a fraction, decimal, percentage, or ratio

posibilidad la probabilidad de una ocurrencia, a menudo expresada como fracción, decimal, porcentaje o razón

circle graph (pie chart) a way of displaying statistical data by dividing a circle into proportionally-sized "slices"

gráfica circular (gráfica de pastel) una forma de demostrar datos estadísticos dividiendo un círculo en "tajadas" proporcionalmente dimensionales

Example/**Ejemplo:**

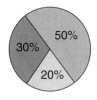

Favorite Color
Color Favorito

circumference the distance around a circle, calculated by multiplying the diameter by the value of pi

circunferencia la distancia alrededor de un círculo calculada mediante la multiplicación del diámetro por el valor de pi

columns vertical lists of numbers or terms; in spreadsheets, the names of cells in a column all beginning with the same letter {A1, A2, A3, A4, . . .}

columnas listas verticales de números o términos; en las hojas de cálculo, los nombres de las casillas en una columna que comienzan con la misma letra {A1, A2, A3, A4, . . . }

congruent figures figures that have the same size and shape. The symbol ≅ is used to indicate congruence.
figuras congruentes figuras que tienen el mismo tamaño y forma. El símbolo ≅ se usa para indicar congruencia.

Example/Ejemplo:

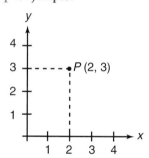

Triangles *ABC* and *DEF* are *congruent*.
Los triángulos *ABC* y *DEF* son *congruentes*.

coordinate graph the representation of points in space in relation to reference lines—usually, a horizontal *x*-axis and a vertical *y*-axis
gráfica de coordenada la representación de puntos en el espacio en relación con las rectas de referencia—generalmente un eje *x* horizontal y un eje *y* vertical

coordinates an ordered pair of numbers that describes a point on a coordinate graph. The first number in the pair represents the point's distance from the origin $(0, 0)$ along the *x*-axis, and the second represents its distance from the origin along the *y*-axis.
coordenadas un par de números ordenados que describen un punto en una gráfica de coordenadas. El primer número en el par representa la distancia del punto desde el origen $(0, 0)$ a lo largo del eje *x* y el segundo representa su distancia desde el origen a lo largo del eje *y*.

Example/Ejemplo:

Point *P* has *coordinates* (2, 3).
El punto *P* tiene las *coordenadas* (2, 3).

counterexample a specific example that proves a general mathematical statement to be false
contraejemplo un ejemplo específico que demuestra la falsedad de un enunciado matemático general

cross product a method used to solve proportions and test whether ratios are equal: $\frac{a}{b} = \frac{c}{d}$ if $ad = bc$
producto cruzado un método usado para resolver proporciones y comprobar si los radios son iguales: $\frac{a}{b} = \frac{c}{d}$ si $ad = bc$

cube (v.) to multiply a number by itself and then by itself again
cubicar multiplicar un número dos veces por sí mismo

Example/Ejemplo: $2^3 = 2 \times 2 \times 2 = 8$

cube root the number that must be multiplied by itself and then by itself again to produce a given number
raíz cúbica el número que se multiplica dos veces por sí mismo para obtener un número determinado

Example/Ejemplo: $\sqrt[3]{8} = 2$

D

decimal system the most commonly used number system, in which whole numbers and fractions are represented using base ten
sistema decimal el sistema numérico más comúnmente usado en el cual los números enteros y las fracciones se representan usando bases de diez

Example: Decimal numbers include 1,230, 1.23, 0.23, and -123.
Ejemplo: Los números decimales incluyen 1,230, 1.23, 0.23 y -123.

discount a deduction made from the regular price of a product or service
descuento deducción hecha del precio regular de un producto o servicio

E

equation a mathematical sentence stating that two expressions are equal
ecuación enunciado matemático que expresa que dos expresiones son iguales

Example/Ejemplo: $3 \times (7 + 8) = 9 \times 5$

equivalent equal in value
equivalente igual en valor

equivalent expressions expressions that always result in the same number, or have the same mathematical meaning for all replacement values of their variables
expresiones equivalentes expresiones que siempre tienen el mismo número como resultado o el mismo significado matemático para todos los valores substitutos de sus variables

Examples/Ejemplos: $\frac{9}{3} + 2 = 10 - 5$
$2x + 3x = 5x$

equivalent fractions fractions that represent the same quotient but have different numerators and denominators
fracciones equivalentes fracciones que representan el mismo cociente, pero tienen numeradores y denominadores diferentes

Example/Ejemplo: $\dfrac{5}{6} = \dfrac{15}{18}$

estimation a value that is the result of an approximation or rough calculation
estimación un valor que es el resultado de una aproximación o cálculo aproximado

even number any whole number that is a multiple of 2 {2, 4, 6, 8, 10, 12, ...}
número par cualquier número entero múltiplo de 2 {2, 4, 6, 8, 10, 12, ...}

expense an amount of money paid; cost
gasto cantidad de dinero pagada; costo

experimental probability a ratio that shows the total number of times the favorable outcome happened to the total number of times the experiment was done
probabilidad experimental una razón que muestra el número total de veces que ocurrió un resultado favorable en el número total de veces que se realizó el experimento

expression a mathematical combination of numbers, variables, and operations
expresión combinación matemática de números, variables y operaciones

Example/Ejemplo: $6x + y^2$

F

factor a number or expression that is multiplied by another to yield a product
factor un número o expresión que se multiplica por otro para obtener un producto

Example: 3 and 11 are *factors* of 33.
Ejemplo: 3 y 11 son *factores* de 33.

flip to "turn over" a shape
invertir "voltear" una figura

Example/Ejemplo:

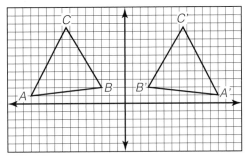

$\triangle A'B'C'$ is a *flip* of $\triangle ABC$.
$\triangle A'B'C'$ es una *inversión* de $\triangle ABC$.

formula an equation that shows the relationship between two or more quantities; a calculation performed by spreadsheet
fórmula una ecuación que muestra la relación entre dos o más cantidades; cálculo realizado en una hoja de cálculo

Example: $A = \pi r^2$ is the *formula* for calculating the area of a circle; A2 × B2 is a spreadsheet *formula*
Ejemplo: $A = \pi r^2$ es la fórmula para calcular el área de un círculo; A2 × B2 es una *fórmula* de una hoja de cálculo

fraction a number representing some part of a whole; a quotient in the form $\dfrac{a}{b}$
fracción un número que representa una parte de un entero; un cociente en la forma $\dfrac{a}{b}$

frequency graph a graph that shows similarities among the results so one can quickly tell what is typical and what is unusual
gráfica de frecuencia una gráfica que muestra similitudes entre los resultados de forma que podamos notar rápidamente lo que es típico y lo que es inusual

H

horizontal a flat, level line or plane
horizontal plano o línea horizontal

I

income the amount of money received for labor, services, or the sale of goods or property
ingreso la cantidad de dinero recibida producto del trabajo, servicios o venta de bienes o propiedad

inequality a statement that uses the symbols > (greater than), < (less than), ≥ (greater than or equal to), and ≤ (less than or equal to) to indicate that one quantity is larger or smaller than another
desigualdad enunciado que utiliza símbolos > (mayor que), < (menor que), ≥ (mayor que o igual a) y ≤ (menor que o igual a) para indicar que una cantidad es mayor o menor que otra

Examples/Ejemplos: $5 > 3$; $\frac{4}{5} < \frac{5}{4}$;
$$2(5 - x) > 3 + 1$$

L

law of large numbers when you experiment by doing something over and over, you get closer and closer to what things "should" be theoretically. For example, when you repeatedly throw a dice, the proportion of 1's that you throw will get closer to $\frac{1}{6}$ (which is the theoretical proportion of 1's in a batch of throws).
ley de los grandes números cuando experimentas haciendo algo varias veces, te acercas cada vez más a cómo "deben" ser las cosas teóricamente. Por ejemplo, cuando tiras un dado repetidamente, la proporción de los 1 que tiras se aproximará a $\frac{1}{6}$ (que es la proporción teórica de los 1 en un grupo de tiradas).

line of symmetry a line along which a figure can be folded so that the two resulting halves match
eje de simetría recta por la que se puede doblar una figura de manera que las dos mitades resultantes sean iguales

Example/Ejemplo:

$\overline{ST}$ is a *line of symmetry.*
$\overline{ST}$ es un *eje de simetría.*

M

mathematical argument a series of logical steps a person might follow to determine whether a statement is correct
argumento matemático una serie de pasos lógicos que una persona podría seguir para determinar si un enunciado es correcto

mean the quotient obtained when the sum of the numbers in a set is divided by the number of addends
media el cociente que se obtiene cuando la suma de los números de un conjunto se divide entre el número de sumandos

Example: The *mean* of 3, 4, 7, and 10 is
$(3 + 4 + 7 + 10) ÷ 4$ or 6.
Ejemplo: La media de 3, 4, 7 y 10 es
$(3 + 4 + 7 + 10) ÷ 4$ o 6.

median the middle number in an ordered set of numbers
mediana el número medio en un conjunto de números ordenado

Example: 1, 3, 9, 16, 22, 25, 27
16 is the *median.*
Ejemplo: 1, 3, 9, 16, 22, 25, 27
16 es la *mediana.*

metric system a decimal system of weights and measurements based on the meter as its unit of length, the kilogram as its unit of mass, and the liter as its unit of capacity
sistema métrico sistema decimal de pesos y medidas que tiene por base el metro como su unidad de longitud, el kilogramo como su unidad de masa y el litro como su unidad de capacidad

N

natural variability the difference in results in a small number of experimental trials from the theoretical probabilities
variabilidad natural la diferencia en los resultados de una pequeña cantidad de ensayos experimentales de probabilidades teóricas

net a two-dimensional plan that can be folded to make a three-dimensional model of a solid
red un diagrama bidimensional que se puede doblar para hacer un modelo tridimensional de un cuerpo sólido

Example/Ejemplo:

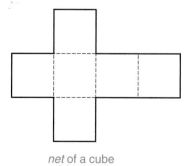

net of a cube
red de un cubo

O

ordered pair two numbers that tell the *x*-coordinate and *y*-coordinate of a point
par ordenado dos números que expresan la coordenada *x* y la coordenada *y* de un punto

> Example: The coordinates (3, 4) are an *ordered pair*. The *x*-coordinate is 3, and the *y*-coordinate is 4.
> Ejemplo: Las coordenadas (3, 4) son un *par ordenado*. La coordenada *x* es 3 y la coordenada *y* es 4.

outcome a possible result in a probability experiment
resultado efecto y consecuencia posible de un experimento de probabilidad

outcome grid a visual model for analyzing and representing theoretical probabilities that shows all the possible outcomes of two independent events
cuadrícula de resultados modelo visual para analizar y representar probabilidades teóricas que muestra todos los posibles resultados de dos sucesos independientes

Example/Ejemplo:

The sample space for rolling a pair of dice
El espacio del modelo donde se lanza un par de dados

	1	2	3	4	5	6
1	(1, 1)	(2, 1)	(3, 1)	(4, 1)	(5, 1)	(6, 1)
2	(1, 2)	(2, 2)	(3, 2)	(4, 2)	(5, 2)	(6, 2)
3	(1, 3)	(2, 3)	(3, 3)	(4, 3)	(5, 3)	(6, 3)
4	(1, 4)	(2, 4)	(3, 4)	(4, 4)	(5, 4)	(6, 4)
5	(1, 5)	(2, 5)	(3, 5)	(4, 5)	(5, 5)	(6, 5)
6	(1, 6)	(2, 6)	(3, 6)	(4, 6)	(5, 6)	(6, 6)

There are 36 possible outcomes. Hay 36 resultados posibles.

P

pattern a regular, repeating design or sequence of shapes or numbers
patrón diseño regular y repetido o secuencia de formas o números

percent a number expressed in relation to 100, represented by the symbol %
por ciento un número expresado con relación a 100, representado por el signo %

> Example: 76 out of 100 students use computers. 76 *percent* of students use computers.
> Ejemplo: 76 de 100 estudiantes usan computadoras. El 76 *por ciento* de los estudiantes usan computadoras.

perfect square a number that is the square of an integer. For example, 25 is a *perfect square* since $25 = 5^2$.
cuadrado perfecto el número que es el cuadrado de un número entero. Por ejemplo, 25 es el cuadrado perfecto como que $25 = 5^2$.

perimeter
the distance around the outside of a closed figure
perímetro la distancia alrededor del contorno de una figura cerrada

Example/Ejemplo:

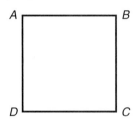

AB + BC + CD + DA = perimeter
AB + BC + CD + DA = perímetro

pi the ratio of a circle's circumference to its diameter. *Pi* is shown by the symbol π, and is approximately equal to 3.14.
pi La razón de la circunferencia de un círculo a la de su diámetro. *Pi* se muestra con el símbolo π y es aproximadamente igual a 3.14.

polygon a simple, closed plane figure, having three or more line segments as sides
polígono una figura plana, simple y cerrada que tiene tres o más líneas rectas como sus lados

Examples/Ejemplos:

polygons
polígonos

price the amount of money or goods asked for or given in exchange for something else
precio cantidad de dinero o bienes pedida o dada a cambio de algo

prime number a whole number greater than 1 whose only factors are 1 and itself
número primo un número entero mayor que 1, cuyos únicos factores son 1 y él mismo.

Examples/Ejemplos: 2, 3, 5, 7, 11

probability the study of likelihood or chance that describes the chances of an event occurring
probabilidad el estudio de las probabilidades que describen las posibilidades de que ocurra un suceso

profit the gain from a business; what is left when the cost of goods and of carrying on the business is subtracted from the amount of money taken in
ganancia lucro de una empresa; lo que queda cuando se resta el costo de los bienes y los gastos que corre la empresa de la cantidad de dinero obtenida

proportion a statement that two ratios are equal
proporción igualdad de dos razones

Q

quadrilateral a polygon that has four sides
cuadrilátero un polígono que tiene cuatro lados

Examples/Ejemplos:

quadrilaterals
cuadriláteros

qualitative graphs a graph with words that describes such things as a general trend of profits, income, and expenses over time. It has no specific numbers.
gráfica cualitativa gráfica de palabras que describe temas como el movimiento general de las ganancias, el ingreso y los gastos en un periodo de tiempo. No tiene números específicos.

R

radius a line segment from the center of a circle to any point on its circumference
radio recta que va desde el centro del círculo hasta cualquier punto de su circunferencia

rate [1] fixed ratio between two things; [2] a comparison of two different kinds of units, for example, miles per hour or dollars per hour
tasa [1] razón fija entre dos cosas; [2] comparación de dos tipos diferentes de unidades, por ejemplo, millas por hora o dólares por hora

ratio a comparison of two numbers
razón una comparación de dos números

Example: The *ratio* of consonants to vowels in the alphabet is 21:5.
Ejemplo: La razón entre las consonantes y las vocales en el abecedario es de 21:5.

rectangle a parallelogram with four right angles
rectángulo paralelogramo con cuatro ángulos rectos

Example/Ejemplo:

rectangle/rectángulo

reflection a type of transformation where a figure is flipped over a line of symmetry
reflexión un tipo de transformación donde una figura se invierte por su eje de simetría

Example/Ejemplo:

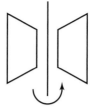

reflection of a trapezoid
reflexión de un trapezoide

rotation a transformation in which a figure is turned a certain number of degrees around a fixed point or line
see turn
rotación una transformación en la cual una figura gira un cierto número de grados alrededor de un punto fijo o eje
ver girar

Example/Ejemplo:

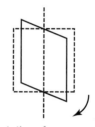

rotation of a square
rotación de un cuadrado

row a horizontal list of numbers or terms. In spreadsheets, the labels of cells in a *row* all end with the same number (A3, B3, C3, D3, . . .)
fila lista horizontal de números o términos. En las hojas de cálculo, los rótulos de las casillas en una fila terminan en el mismo número (A3, B3, C3, D3, . . .)

S

scale the ratio between the actual size of an object and a proportional representation
escala la razón entre el tamaño real de un objeto y una representación proporcional

scale size the proportional size of an enlarged or reduced representation of an object or area
tamaño de escala el tamaño proporcional de una representación aumentada o reducida de un objeto o área

scatter plot (or scatter diagram) a two-dimensional graph in which the points corresponding to two related factors (for example, smoking and life expectancy) are graphed and observed for correlation
gráfica de dispersión (diagrama de dispersión) gráfica bidimensional donde los puntos correspondientes a dos factores relacionados (por ejemplo, fumar y la esperanza de vida) se representan gráficamente y se observan para su correlación

Example/Ejemplo:

AGE AND DIAMETER OF RED MAPLE TREES
EDAD Y DIÁMETRO DE LOS
ÁRBOLES ARCE ROJOS

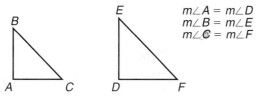

signed number a number preceded by a positive or negative sign. Positive numbers are usually written without a sign.
número con signo un número precedido por un signo positivo o negativo. Los números positivos por lo general se escriben sin signo.

similar figures have the same shape but are not necessarily the same size
figuras similares que tienen la misma forma, pero no necesariamente el mismo tamaño

Example/Ejemplo:

$m\angle A = m\angle D$
$m\angle B = m\angle E$
$m\angle C = m\angle F$

Triangles *ABC* and *DEF* are *similar figures*.
Los triángulos *ABC* y *DEF* son *figuras similares*.

simulation a mathematical experiment that approximates real-world process
simulación experimento matemático que aproxima el proceso del mundo real

slide to move a shape to another position without rotating or reflecting it
desplazar mover una figura hacia otra posición sin rotarla o reflejarla

Example/Ejemplo:

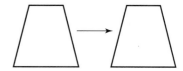

slide of a trapezoid
desplazamiento de un trapezoide

solution the answer to a mathematical problem. In algebra, a *solution* usually consists of a value or set of values for a variable.
solución la respuesta a un problema matemático. En álgebra, una *solución* generalmente consiste en el valor o conjunto de valores de una variable.

spreadsheet a computer tool where information is arranged into cells within a grid and calculations are performed within the cells. When one cell is changed, all other cells that depend on it automatically change.
hoja de cálculo herramienta de las computadoras donde la información es organizada en casillas dentro de una cuadrícula y se realizan cálculos dentro de las casillas. Cuando una casilla cambia, el resto de las casillas que dependen de ella cambian automáticamente.

square number a number that can be expressed as x^2
número cuadrado un número que se puede expresar como x^2

Examples/Ejemplos: 1, 4, 9, 16, 25, 36

square root a number that when multiplied by itself produces a given number. For example, 3 is the *square root* of 9.
raíz cuadrada un número que cuando se multiplica por sí mismo produce un número determinado. Por ejemplo, 3 es la *raíz cuadrada* de 9.

Example/Ejemplo: $3 \times 3 = 9$; $\sqrt{9} = 3$

statistics the branch of mathematics concerning the collection and analysis of data
estadística la rama de las matemáticas que estudia la colección y análisis de datos

strip graph a graph indicating the sequence of outcomes. A *strip graph* helps to highlight the differences among individual results and provides a strong visual representation of the concept of randomness.
gráfica de franjas gráfica que indica la secuencia de resultados. Una gráfica de franjas ayuda a destacar las diferencias entre resultados individuales y provee una representación visual fuerte del concepto de aleatoriedad

Example/Ejemplo:

Outcomes of a coin toss
Resultados de un lanzamiento de monedas
H = heads T = tails
H = cara   T = cruz

strip graph
gráfica de franjas

symmetry *see line of symmetry*
simetría *ver eje de simetría*

Example/Ejemplo:

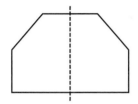

This hexagon has *symmetry* around the dotted line.
Este hexágono tiene *simetría* alrededor de la línea de puntos.

T

table a collection of data arranged so that information can be easily seen
tabla una colección de datos organizados de forma que la información pueda ser vista fácilmente

theoretical probability the ratio of the number of favorable outcomes to the total number of possible outcomes
probabilidad teórica la razón del número de resultados favorables en el número total de resultados posibles

three-dimensional having three measurable qualities: length, height, and width
tridimensional que tiene tres propiedades de medición: longitud, altura y ancho

transformation a mathematical process that changes the shape or position of a geometric figure *see reflection, rotation, translation*
transformación proceso matemático que cambia la forma o posición de una figura geométrica *ver reflexión, rotación y traslación*

translation a transformation in which a geometric figure is slid to another position without rotation or reflection
traslación transformación en la que una figura geométrica se desplaza hacia otra posición sin rotación o reflexión

trapezoid a quadrilateral with only one pair of parallel sides
trapezoide un cuadrilátero con un solo par de lados paralelos

Example/Ejemplo:

trapezoid
trapezoide

trend a consistent change over time in the statistical data representing a particular population
tendencia cambio consistente con en el tiempo de los datos estadísticos que representan a una población en particular

triangle a polygon that has three sides
triángulo un polígono que tiene tres lados

turn to move a geometric figure by rotating it around a point
girar mover una figura geométrica rotándola a su alrededor

Example/Ejemplo:

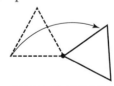

turning of a triangle
giro de un triángulo

two-dimensional having two measurable qualities: length and width
bidimensional que tiene dos propiedades de medición: longitud y ancho

U

unit cost the cost of an item per unit, such as *per ounce* or *each*
costo de unidad el costo de un artículo por unidad, tal como *por onza* o *de cada uno*

unit price the price of an item expressed in a standard measure, such as *per ounce* or *per pint* or *each*
precio de unidad el precio de un artículo expresado en su medida estándar, tal como *por onza* o *por pinta* o *de cada uno*

V

vertex (pl. *vertices*) the common point of two rays of an angle, two sides of a polygon, or three or more faces of a polyhedron
vértice el punto común de las dos semirrectas de un ángulo, dos lados de un polígono o tres o más caras de un poliedro

Examples/**Ejemplos:**

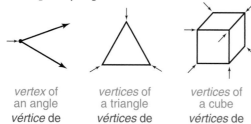

vertex of
an angle
vértice de
un ángulo

vertices of
a triangle
vértices de
un triángulo

vertices of
a cube
vértices de
un cubo

vertical a line that is perpendicular to a horizontal base line
vertical recta que es perpendicular a la recta horizontal

Example/**Ejemplo:**

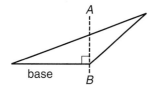

$\overline{AB}$ is *vertical* to the base of this triangle.
$\overline{AB}$ es *vertical* a la base de este triángulo.

W

what-if questions a question posed to frame, guide, and extend a problem
preguntas de "¿Qué sucedería si..?" interrogación hecha para formular, dirigir y ampliar un problema

X

x-axis the horizontal reference line in the coordinate graph
eje x recta de referencia horizontal en la gráfica de coordenadas

Y

y-axis the vertical reference line in the coordinate graph
eje y recta de referencia vertical en la gráfica de coordenadas

INDEX

P

Parallel lines, 284–285, 306

Parallelogram
properties of, 284–285, 306

Patterns
area and cost, 160–161, 166–167
diameter and circumference of a
circle, 294–295
divisibility, 116–117, 132
in the equations of horizontal and
vertical lines, 198, 218
exponential, 112–113
Fibonacci sequence, 86
lines of symmetry for regular
polygons, 290–291
on a multiplication chart, 120–121,
134
in the ordered pairs for an equation,
195, 216
perfect square, 106–107, 128
perimeter, 219
polygon sides and angles, 286–287,
307
predicting with, 190–191, 200,
210–211, 215, 219, 223
prime number, 118–119, 133
quadrilateral sides and angles,
286–287, 307
triangle sides and angles, 276–277

Percent, 24–33, 42–45
budgets and, 32–33, 45
circle graphs and, 28–29, 32–33, 43
cost estimation and, 162–163, 177
discount savings, 30–31, 44
estimating, 26–27, 30, 42
mental calculation, 26–27, 42
of a number, 162–163, 177
probability expressed as, 54–55,
60–63, 70–71, 82, 84, 85, 88
proportion and, 162–163

Perimeter
of an irregular figure, 144–145, 170
patterns, 219
of a regular polygon, 297

Pi, 294–295, 310

Polygon, 282–291, 306–309
angle sums, 286–287, 307
apothem, 297
area, 160–161, 296–297, 310
classification, 150, 284–285, 306
comparison, 291
concave, 306
congruent, 288–289, 291, 308
convex, 306
definition, 284
hexagons, 219
perimeter, 219, 297
regular, 291, 297
vertex, 296

Polyhedron. *See also* three-dimensional
figures
classification, 150–153
surface area, 154–155

Power, 112–113

Prediction
using divisibility, 120–121
using equations, 215
using equations, tables and graphs,
210–211, 223
using examples and counter
examples, 98–99
using experimental results, 50–53,
80, 81
using linear graphs, 200, 211, 219, 223
using patterns, 106–107, 128
using proportion, 23, 68, 70–71, 88
using qualitative graphs, 266
using survey data, 244–245
using what-if questions, 254–255

Price
graphing, 8–9, 35
income, profit relationship, 246–247,
265
sales relationship, 244–245, 264
unit price and, 6–13, 34–37

Price graph
creating, 9, 35
interpreting, 8, 35

Prime factor, 118–119, 133

Prime number, 86, 118–119, 133

Prism, 150–153, 172, 173

Probability
area models, 56–63, 83–85
comparing theoretical and
experimental, 66–71, 86–88
event, 54
experimental, 50–55, 66–71, 80–82,
86–88
expressed as fractions, decimals, and
percents, 53–55, 60–63, 70–71, 82,
84, 85, 88
fair and unfair games, 64–71, 86–88
impossible event, 54
independent event, 66–69, 86, 87
law of large numbers, 52–53, 81
likelihood of events, 51, 54–55
outcome grids, 66–71, 86–88
ratios, 54–55, 60–63, 70–71, 82, 84,
85, 88
simulations and, 72–79, 89–91
theoretical, 53–55, 66–71, 82, 86–88
variability of results, 50–53, 80, 81
words and numbers for, 60–61, 84

Probability line, 54–55, 76–77, 82, 89

Profit
calculation on a spreadsheet,
240–241, 263
improvement study, 252–257,
267–269
income, expense calculation,
230–231, 259
income, expense formula, 232–233,
260
income, expense relationships,
228–229, 256–257, 258, 269
income, expense simulation,
230–233

price, income, relationships,
246–247, 265

Properties
algebraic, 188–189, 214
integer, 102–103, 127
polygon, 238–239, 260

Proportion, 20–23, 40, 41
cross products, and, 21
definition, 15
equal ratios and, 20–21, 40
percent and, 162–163
predicting outcomes with, 70–71, 88
scale and, 39, 139, 140–145,
168–170
setting up, 22–23, 41
ways to solve, 20

Protractor
angle measurement, 274–279,
286–287, 302
circle graph construction, 28–29, 43

Puzzles
integer, 126, 127

Pyramid, 150–153, 172, 173

Q

Quadrants, of the coordinate plane,
194

Quadrilateral
angle sum, 286–287, 307
classification, 284–285, 306

Qualitative graphs, 248–249, 266

R

Radius, 294–295, 310

Ranking
events by likelihood, 76–77, 89
games by probability of winning,
70–71, 88
probabilities, 54–55, 82

Rate
price graph of, 8–9, 35
total price and, 12–13, 37
unit price, 4–13, 34–37

Ratio, 16–19, 38, 39 *See also*
Percent; Rate
comparison, 17, 38
cross products and, 21
definition, 15
equal, 18–19, 38
pi, 294–295, 310
probabilities expressed as, 54–55,
60–63, 70–71, 82, 84, 85, 88
proportion and, 20–23, 40, 41
scale and, 39, 139, 140–145,
168–170
table, 16

PHOTO CREDITS